Ad Hoc Networks

Ad Hoc Networks

Routing, QoS and Optimization

Mounir Frikha

First published 2011 in Great Britain and the United States by ISTE Ltd and John Wiley & Sons, Inc. Adapted and updated from *Réseaux ad hoc : routage, qualité de service et optimisation* published 2010 in France by Hermes Science/Lavoisier © LAVOISIER 2010

ISTE Ltd
27-37 St George's Road
London SW19 4EU
UK

www.iste.co.uk

John Wiley & Sons, Inc.
111 River Street
Hoboken, NJ 07030
USA

www.wiley.com

Library of Congress Cataloging-in-Publication Data

Frikha, Mounir.
Ad hoc networks : routing, quality of service, and optimization / Mounir Frikha.
p. cm.
Includes bibliographical references and index.
ISBN 978-1-84821-227-5 (hardback)
1. Ad hoc networks (Computer networks) I. Title.
TK5105.77.F75 2011
004.6'8--dc22

2010042678

British Library Cataloguing-in-Publication Data
A CIP record for this book is available from the British Library
ISBN 978-1-84821-227-5

Printed and bound in Great Britain by CPI Antony Rowe, Chippenham and Eastbourne.

Table of Contents

Chapter 1

Introduction to Ad Hoc Networks

1.1. Introduction

With recent technological advances in the domain of wireless communications and the emergence of portable computing devices, researchers have turned their attention to improving the function of networks and, in particular, to ensuring rapid access to information independent of time or place.

Until very recently, wireless networks were based exclusively upon planned and sized infrastructures and a hierarchical control of operations. With the vast expansion of wireless applications, particularly personal and local networks, the need for self-organization, independence, adaptability, and cost reduction has become apparent. Mobile ad hoc networks (MANETs) provide a solution that effectively responds to these issues.

The concept of MANETs attempts to expand notions of mobility to all parts of the environment. Unlike networks based on cellular communication, no central administration is available; mobile operators themselves provide a network infrastructure on an ad hoc basis. No supposition or limitation is made regarding the size of the ad hoc network, which can contain hundreds, or even thousands, of cellular units.

In this chapter, we present the evolution from networks with fixed infrastructures to MANET networks. Next, we list the different characteristics of ad hoc networks and give a few examples of their application.

1.2. Wireless networks and communications

In the past few decades, technological advancements have made major changes to wireless networks, from the techniques used for transmission to the nature of services offered.

With the study of the development of these networks in mind, we begin by presenting the characteristics inherent in wireless communications.

1.2.1. *Wireless communications*

In general terms, the expression "wireless communications" refers to "communications involving infra-red signals or radio frequencies and allowing the exchange of information and resources between the different entities of a network" [BAS 04].

Wireless communications vary according to the range and the type of modulation used.

1.2.2. *Wireless networks*

By the nature of the transmission channel used, wireless networks are distinguished from wired networks by a number of characteristics [BAS 04]:

– *An unpredictable environment*. Interference, mobility, changing channels, and variations in the strength of the signal are all factors that make the network extremely variable.

– *Unreliable medium*. Transmission over a radio channel is prone to errors. Furthermore, interference and the unpredictable quality of links reduce the reliability of the medium. In addition, limited capacity means that the protocols in the transport layer responsible for reliability may not be supported by network nodes.

– *Limited resources*. In the case of mobile nodes, power is supplied by batteries. For reasons of weight and practicality, these nodes have limited storage capacity and limited processing power. Finally, radio channels are a scarce and costly shared resource for which the usage is governed by restrictive regulations:

- the limited size of equipment/tools due to the demands of portability,

- diminished security as the radio interface is shared, and

- dynamic topology.

Wireless networks are considerably more dynamic in nature than wired networks. This is particularly true in the case of mobile networks. From the moment nodes are able to move in and out of the range of other nodes, connections within the network can be cut and others can form.

1.2.3. *Classification of wireless networks*

Wireless networks can be categorized using various classifications, according to the size of the zone covered, the architecture of the network, or the technique used for access to the radio channel.

1.2.3.1. *Classification by type of network architecture*

We can distinguish three classes of wireless network according to the level of communication infrastructures used within the network:

– *Networks entirely dependent on communication infrastructures*. In this type of network, a node can only access the network via a communication infrastructure deployed by the network. This infrastructure could be an access point, a wireless bridge, a wireless access router, or a base station transceiver, among other things. The type of network access infrastructure is dependent on various parameters, including the type of application of the network, the range of the network, the envisaged coverage, and the mobility of the nodes.

– *Networks without a communication infrastructure*. From the Latin meaning "which goes where it must," ad hoc networks are formed dynamically by the cooperation of a random number of independent nodes. The role of each node is not predetermined and nodes make decisions dependant on the situation of the network without having recourse to a preexisting infrastructure. For example, two PCs equipped with wireless network cards can form an ad hoc network each time one comes into the signal range of the other. As no communication infrastructure is used, a "hopping" technique is used to access the network. In other words, to communicate with a destination, a mobile node makes use of the other nodes in the network, the message being passed from peer to peer.

– *Networks making partial usage of communication infrastructures*. This type of network is known as a grid network and uses communication infrastructures to allow the wireless network to connect to the Internet. Unlike those networks that use infrastructures exclusively, grid networks do not require a communication infrastructure to be within immediate range of each node. If the node does not detect an infrastructure in its immediate vicinity, it uses the hopping technique to reach one. This architecture combines the two methods of deployment previously discussed and the cost of the communication infrastructure is reduced. It also limits the problems inherent in the usage of hopping. However, problems of overloading can occur in the node immediately next to the communications infrastructure as it receives the accumulated traffic of several of the nodes in the network that it must transmit to this infrastructure.

1.2.3.2. *Classification by extent of the zone covered*

Using a system of classification by extent of the zone covered by a wireless network, we can distinguish four types of network:

– *Wireless wide area networks* (*WANs*). Wireless WANs are networks built on infrastructures of which the range can cover a wide geographical area – whole towns or even countries. Communications are transmitted via multiple antennae or by satellite. Cell networks, for example GSM, and satellite networks fall into this category.

– *Wireless metropolitan area networks.* These metropolitan networks are sometimes described as “fixed” wireless networks. Like WAN, they use an infrastructure and allow wide-bandwidth connections between sites in the same metropolitan zone. University campus networks or networks linking neighboring residential blocks are good examples of this type of network.

– *Wireless local area networks* (*WLAN*). These local wireless networks allow wireless communications to be set up within an establishment, for example, business premises or an airport. WLAN networks can operate either with an infrastructure or on an ad hoc basis. In cases where an infrastructure is used, stations connect to linked access points that permit interface with the wired backbone. In an ad hoc network, stations within a limited zone, for example, a conference room, can form a temporary network without recourse to access points.

– *Wireless personal area networks* (*WPANs*). WPANs allow users to communicate over a short distance between personal wireless devices, for example, PDAs, mobile telephones, or laptop computers, within a very limited area – the maximum radius is around 10 m [BAS 04]. Two key technologies are used in PAN networks: Bluetooth and Infrared. Bluetooth technology is used in the place of wired connections to transmit data over a maximum distance of 10 m. Infrared has a maximum range of 1 m [BAS 04].

WPANs have enjoyed a high degree of success due to its simplicity, low energy consumption, and interoperability with IEEE 802.11.

1.2.3.3. *Classification by means of access to the radio channel*

By their very nature, radio channels are shared channels, and it has been necessary to find adequate means of access to organize equitable sharing of resources while reducing interference between the different communications sharing the channel as far as possible.

1.2.3.3.1. Time division multiple access (TDMA) networks

TDMA is a means of multiplexing communications that consists of dividing the time available between different users. For example, in GSM, a carrier frequency is divided into eight time slots, thus enabling the network to carry up to eight communications simultaneously.

The use of TDMA techniques has the advantage of permitting simultaneous usage of the same frequency by multiple users, allowing an advantageous exploitation of network resources (as frequency band is a scarce and costly resource).

1.2.3.3.2. Frequency division multiple access (FDMA) networks

FDMA is a method of access to a radio channel based on sharing a frequency band between a number of simultaneous communications. In GSM, each wave band of 25 MHz is divided between 124 subcarriers.

This technique has the advantage of allowing a communication without temporary interruptions, as a user may communicate during every single moment of the available time. However, the division of a frequency between multiple users is not possible within the same radio cell, and the management of radio resources is therefore not optimized.

1.2.3.3.3. Code division multiplex access (CDMA) networks

The CDMA method allows simultaneous shared access to a radio channel without sharing either time or the wave band between different users. The principle of CDMA is to give each communication a unique random sequence of pseudo-codes, which allows each communication to be separated from external interference upon reception at the point of decoding.

In exploiting the full available frequency band, CDMA also has the advantage of using the spread spectrum of the radio signal, making it possible to avoid interference and scrambling at particular frequencies and rendering communications more secure.

1.2.3.3.4. Space division multiplex access (SDMA) networks

In mobile networks, the base station has no information concerning the position of a mobile node. This infers a loss of transmission power in the antenna, as the signal is emitted over the whole zone covered by the antenna, whereas the mobile station is only available in one direction. In addition, this method creates interference problems for neighboring cells.

The SDMA technique combines the usage of the so-called smart antennas with the exploitation of information concerning the localization of the mobile unit to adapt the radiation pattern of the antenna accordingly.

1.3. Ad hoc networks (MANET)

MANET is the name of an Internet engineering task force (IETF) work group created in 1998 with the aim of standardizing routing protocols based on Internet protocol technology for ad hoc networks, mobile, etc.

Since the inception of this work group, the term MANET has come to be used to describe any ad hoc mobile network.

In general, a MANET network is created in a dynamic manner by an autonomous system of mobile nodes connected via wireless links, without the need for an external infrastructure or a centralized administration.

To do this, a "hopping" technique is used. The operation of a MANET network thus depends on the collaboration of all elements of the network in routing.

The concept of ad hoc networks has been the subject of scientific research since 1970s. A true definition of an ad hoc network is hard to pin down. In existing literature on the subject, the term is used in different ways. The IETF, the body responsible for guiding the evolution of the Internet, gives the following definition:

An ad hoc mobile network (MANET) is an autonomous system of mobile routers (and associated hosts) connected by wireless links. These routers are free to move at random and organise themselves arbitrarily; thus, the topology of the network can change rapidly and in an unpredictable manner. A network of this type can operate in autonomy or be connected to the Internet. MANETs are used in many applications as they do not require any infrastructure support.

MANETs are made up of mobile hosts connected by radio links without possessing a fixed infrastructure or a central administration.

Communication is carried out directly between nodes or via intermediary nodes acting as routers.

The principle advantage of this type of network is the velocity with which it can be deployed, its robustness, its flexibility, and its ability to deal with node mobility.

In certain environments where MANET networks are used, for example, military communications, national crises, or natural disasters, wired networks are not available; in such cases, ad hoc networks are the only available means of communication and access to information.

Furthermore, ad hoc networks play an important role in civil forums such as campus recreations, conferences, and electronic classes.

Ad hoc networks can be considered as a wireless Internet network where users are free to move geographically while retaining the possibility of connection with the rest of the world.

The successful implementation of ad hoc wireless network technology presents a challenge in which the implementation of this kind of network makes different demands to those of traditional wireless systems or wired networks.

An ad hoc network can be modeled on a graph as $Gt = (Vt, Et)$, where Vt represents the totality of the nodes (mobile units or hosts) present in the network and Et shows all existing connections between the nodes (Figure 1.1). If $e = (u,v) \in Et$, it signifies that the nodes u and v are able to communicate directly at that moment.

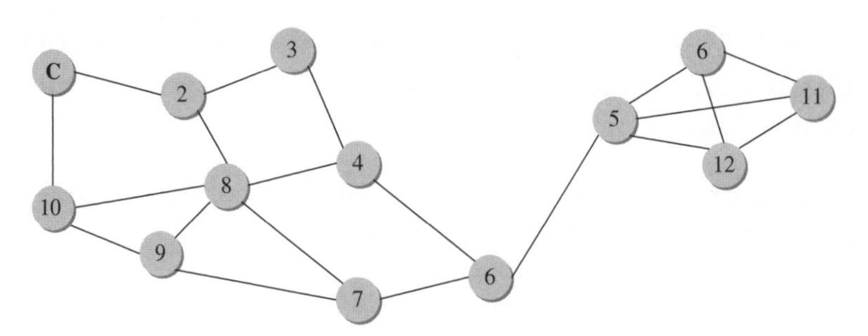

Figure 1.1. *Model of an ad hoc network*

A network is frequently subjected to topological changes because of the mobility of its component nodes. These topological changes impact on the distribution of connections between different nodes. In networks of this kind, radio links are broken on a regular and unpredictable basis. Figure 1.2 illustrates this effect where node 1, at a given moment $t + 2$, disappears from the network as the link between node 1 and node 4 ceases to exist.

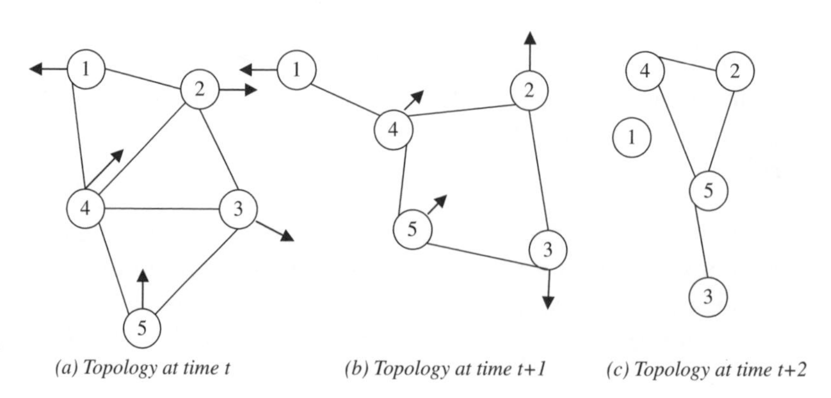

(a) Topology at time t
(b) Topology at time t+1
(c) Topology at time t+2

Figure 1.2. *Illustration of mobility in an ad hoc network*

1.3.1. *Characteristics and advantages*

Ad hoc networks can be used in any external environment. In this respect, the deployment of an ad hoc network is simple: the only requirement is a certain number of terminals in a given space. It is also rapid, becoming operational the moment all terminals are present.

Certain other characteristics are typical of networks of this nature:

– *Absence of infrastructure*. Ad hoc networks can be distinguished from other types of mobile network by the absence of a preexisting infrastructure and of any form of centralized administration. Mobile hosts are responsible for the establishment and maintenance of connectivity over the network on a continuous basis.

– *Dynamic topology*. The mobile units that make up the network move freely and arbitrarily, with the ability to join or leave the network at any time. As a result, the network topology can change rapidly and in a random manner at any time. Within the network topology, links can be uni- or bi-directional.

– *Limited bandwidth, variable capacity of links*. Bandwidth is restricted in comparison with that offered by a wired network as the wireless transmission channel is shared. Congestion is a major consequence of limited bandwidth, a problem not helped by diffusion. In fact, every diffusion packet transmitted to a station already in the course of communication (whether the packet is destined for that station or not) will alter that station's communication.

– *Energy constraints*. Mobile hosts are powered by autonomous energy sources, for example, batteries or other consumable sources. The energy parameter is relatively important as it affects the life expectancy of the network: nodes are required to adjust their energy consumption to reach neighbors or a destination. For this reason, the energy parameter must be taken into account in all monitoring undertaken by the system. A number of research groups are currently occupied in attempting to improve the energetic performance of these systems without adverse effects on their applications.

– *Limited security*. MANETs are more affected by security parameters than classic wired networks. In the latter case, a network can be secured easily, thanks to the presence of a central administration. Conversely, ad hoc networks do not contain a node with the role of administrator. The security systems used by wired networks are therefore not adapted to use in an ad hoc network. To compensate for this problem, monitoring of transferred data must be minimized to preserve both energy and resources.

1.3.2. *Applications*

The flexibility and distributed architecture of ad hoc networks give them considerable commercial potential and make them very easy to put into place.

From a historical point of view, MANET networks were first used by the military as part of tactical networks to improve battlefield communications. The nature of military operations made the use of a fixed communication infrastructure difficult in combat situations.

Moreover, conventional wireless communications present limitations due to the different extenuations undergone by radio waves in the course of their propagation. Radio signals above 100 MHz rarely travel beyond the line of sight (LOS) [BAS 04].

MANET networks offer a user-friendly solution to these problems, mainly thanks to their use of the "hopping" technique. They provide a distributed architecture without infrastructure while permitting connectivity over and beyond the LOS.

The first application of MANET networks dates back to the DARPA pocket radio network (PRNet) project in 1972 [BAS 04]. In a period marked by the appearance of methods of shared access to media (for example, CSMA and ALOHA) and methods using packet-switching (store-and-forward), the PRNet project aimed to use these mechanisms in mobile wireless networks for data transmission.

Subsequently, another DARPA project, the survivable radio networks project launched in 1983, considered issues of scalability, security, treatment capacity, and power management. The core work undertaken on this project was improved using digital low-cost packet radio technology in 1987 [BAS 04] using the direct sequence spread spectrum technique.

Although MANET networks were initially used mostly in military contexts, a number of civil applications of ad hoc networks have emerged, particularly since the appearance of the Internet. Ad hoc networks are now used for a variety of local and personal network applications. To cite some examples [BAS 04], ad hoc networks are now used for connections between computers, PDA devices, civil policing applications, rescue operations, emergency services, communication between vehicles, networks of sensors, domestic robots, and multiplayer games, among other uses.

Although projects concerning wireless networking in general and ad hoc networks more particularly started out in a purely military framework (Figure 1.3), their domains of application now expand well outside the military sphere. Wireless networks offer a high degree of flexibility and rapidity while being simple to implement. They would be of great value in the cases of natural disaster or fire, situations where the rapid implementation of a network would be essential to coordinate the efforts of the emergency services and rescue operations.

Figure 1.3. *Use of ad hoc networks in military operations*

Wireless networks are easier to put into place than traditional wired connections in buildings where it would be difficult or undesirable to install a suitable system of cables, for example, old or listed buildings or for large-scale sporting events.

This kind of network can also be used to inter-connect buildings when cost is an issue as it is not necessary to rent a specialist link from an operator.

Industrial applications of ad hoc network technology can also be envisaged; for example, a network could be installed in a factory where the nodes would be mobile robots, free to move within the factory while maintaining communications.

Another possible application would be in hostile environments, such as the crater of a volcano, where an ad hoc network could be used to monitor volcanic activity, or, to give another example, along a fault line.

On a university campus, the use of wireless networks would potentially be very useful for students, who would be able to connect to their accounts and work from the library or, indeed, their rooms.

The need for ad hoc networks is becoming more and more apparent, notably in the domain of traffic management. The use of a network of this kind permits messages concerning the state of the roads to be transmitted, raising awareness of obstructions, for example (Figure 1.4).

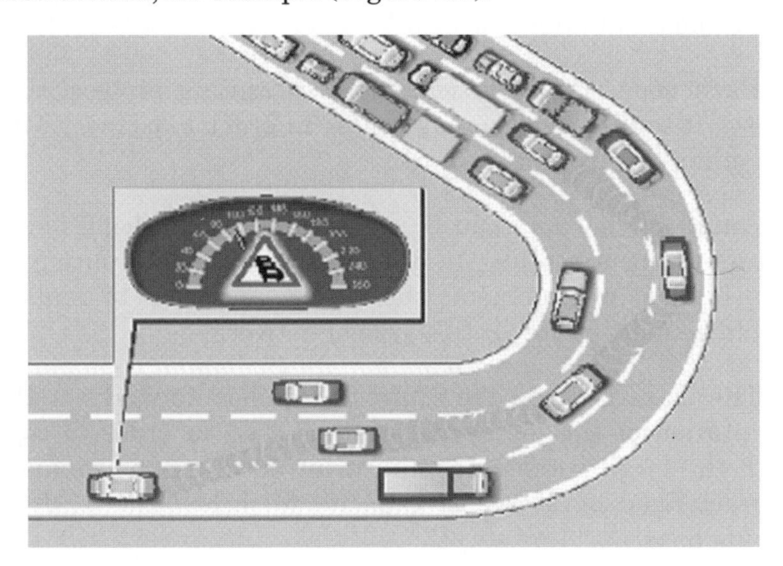

Figure 1.4. *Use of ad hoc networks on the road*

1.4. Routing of ad hoc networks

Generally speaking, routing is a method of transmitting information to the correct destination across a given network. Routing in an ad hoc network is based on a simple and intuitive approach: the re-emission of packets by each node allows propagation within the network. The problem lies in the choice of an optimal route. This essentially comes down to calculating the "best" route to join any two given nodes in a network. However, there is always some difficulty in the choice of criteria used to decide whether one route is better than another. A fixed set of rules must be applied to the routing algorithms such that the problem is transformed into a simple search for the shortest route from source to destination.

With the aim of maintaining network connectivity despite the absence of a fixed infrastructure and the mobility of stations, each node may be pressed into service to take part in routing and to retransmit packets from a node which is unable to reach its destination; each node thus plays the role of station and router.

Each node, then, participates in a routing protocol that allows it to discover existing routes in order to connect with the other nodes of the network.

An ad hoc network can be enormous, an additional factor necessitating a completely different approach to routing in this kind of environment; the classic approach to routing would be inadequate for this type of network.

The main issue in the context of ad hoc networks is the adaptation of the method of routing used in order to cope with the large number of units involved in an environment characterized by a modest capacity for calculation and for saving data.

Ad hoc-routing protocols can be classified in various ways: by the method used to create and maintain routes, or according to the way they transmit data.

1.4.1. *Hierarchical routing, flat routing and routing by geographic localization*

Routing protocols for ad hoc networks can be classified using a number of criteria. The first of these concerns the way they envision the network and the roles attributed to different mobiles.

"Flat" routing protocols. These protocols consider every node to be equal (Figure 1.5). The decision of a node to route packets via another depends on its position and can be re-evaluated with the passage of time.

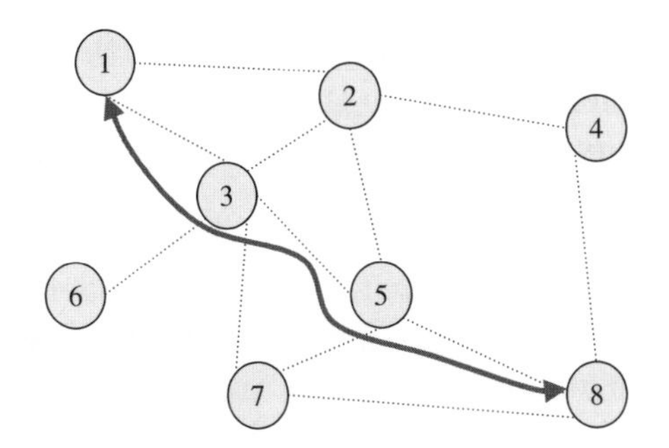

Figure 1.5. *Flat routing*

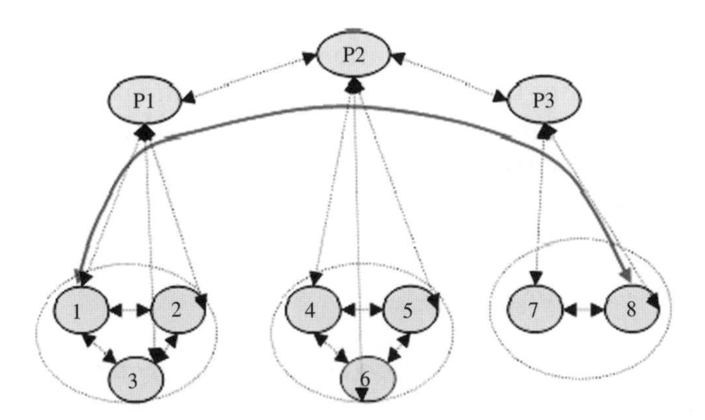

Figure 1.6. *Hierarchical routing*

Hierarchical routing protocols. These protocols operate by giving variable roles to different mobiles. Certain selected nodes assume particular functions, creating a polyvalent vision of the network topology. For example, one mobile might operate as a gateway for a given number of nodes that would be attached to it. This makes routing a simpler affair, with data being passed from gateway to gateway until the gateway linked to the destination node is reached.

Figure 1.6 illustrates an example of hierarchical routing. Node 1 passes through gateways P1, P2, and P3 to reach node 7. In this type of protocol, the gateways do most of the work of routing; mobiles within the network know that if the target destination is not in their neighborhood, they must simply send the data to the relevant gateway, which will take care of it.

In networks where certain nodes are sedentary and have access to sufficient energy (for example, a network of laptop computers where some machines are connected to the mains and/or base stations are established to guarantee connectivity), this form of routing presents certain advantages.

Routing protocols with geographic localization. The advantage of global localization protocols (*global position*) lies in the ability to determine the geographic position of the mobile. Global position system can supply this information within a few metres of accuracy. It also provides universal settings. Satellites are generally used for this purpose and any problem affecting the satellites has knock-on effects on the network.

1.4.2. *Link-state, distance-vector and source-routing protocols*

Another means of classification involves the division of routing protocols into three broad groups: protocols based on link state, those based on distance-vectors, and those that use source-routing.

Link-state protocols. These protocols are based on collected information concerning the state of links within the network (the state of links is typically the group of links between a node and its neighbors). This information is disseminated across the network periodically, allowing each node to build up a map, or graph, of the whole network.

The best-known routing protocol of this type is open shortest path first (OSPF). OSPF is based on Dijkstra's SPF algorithm for the calculation of the shortest route between a source node and the other nodes of the network. One advantage of this type of protocol is the capacity to find alternative routes with ease when a link is broken. It is even possible to use several routes to a same destination simultaneously, thus sharing the workload more effectively and increasing the system's resistance to faults. However, if the network is extensive, the quantity of information to stock and diffuse can become quite considerable.

Link-state protocols can calculate multiple routes, an important attribute when sharing out traffic, for instance, or when replacing a damaged route. As calculations are carried out in a distributed manner at the level of each router, attention must be paid to ensure consistency. Each router must possess the same network information and use the same criteria when faced with multiple choices.

Distance-vector protocols. Protocols in this group work by exchanging information concerning the distance to known destinations between neighboring nodes. Each node sends a list of destinations it can access, with the corresponding costs, to neighboring nodes. The receiving node uses this information to update its local list of destinations with the minimum cost for each. The best-known protocol of this family is undoubtedly the routing information protocol (RIP). The calculation of routes is based on the distributed Bellman–Ford algorithm.

Distance-vector routing protocols have one great advantage in which they are simple to program. However, they are subject to two major problems in the form of loop generation and counting to infinity.

From time to time, nodes can become isolated from the rest of the network when a loop develops. Each node believes that the shortest route to the rest of the network passes through another node than itself. In this case, each exchange of distance-vectors leads to an increase in the perceived cost of transmission. This incremental increase will only stop once a predefined maximum value is reached. This maximum value is set at infinity. In the meantime, data packets may become trapped in loops within the network. Link-state routing protocols can easily avoid loops and are able to make use of multiple selection criteria, as the information exchanged concerns the links in the network: flow, velocity, cost, and reliability.

Source-routing protocols. For this group of protocols, the source of data sets out the exact sequence of nodes a packet will travel through to reach a destination. To send a data packet to another node, the sender creates a source route, which it includes at the beginning of the packet. The route is established by specifying the address of every node the packet must travel through. The sender then transmits the packet using its interface to the first node specified in the source route. A node, upon receiving a packet for which it is not the final destination, removes its own address from the top of the packet and sends the packet on to the next given address. This process is repeated until the packet reaches its final destination. Finally, the packet is delivered to the network layer of the last host.

1.4.3. *Proactive, reactive and hybrid routing*

In work carried out by the IETF, several distinct families of protocols were rapidly identified. Each protocol can be classified as reactive, proactive, or hybrid:

– *Proactive (table-driven) protocols.* These protocols are the successors of distance-vector and link-state protocols. They maintain routing tables containing information on the topology of the network. Each change in network topology triggers updates in the network to maintain an up-to-date global image of the network. The disadvantage of this kind of protocol is in signalization, which can have an effect on bandwidth. The advantage is that a route from source to destination is always available without recourse to route-seeking mechanisms. However, proactive routing protocols have certain shortcomings in cases of frequent and rapid topological development within the network, or when the network is relatively large. An example of this type of protocol is the destination-sequenced distance-vector (DSDV) protocol.

– *Reactive or on demand protocols.* These protocols are the most recent offering in the search for reliable routing solutions in wireless networks. Their defining characteristic is that they carry out a route search when a source wants to communicate with a destination and does not know how to reach it. Route discovery is carried out by "flooding" messages: the source node, seeking a route to a destination, broadcasts a request for information across the network. Upon receiving the request, intermediary (or transit) nodes attempt to teach the source node the route and save the route in the sent table. Once the destination node has been reached, it is able to respond using the route traced by the request, thus establishing a full duplex route between source and destination nodes. The effort involved is reduced in cases where a transit node already possesses a route to the destination. Once the route is calculated, it must be saved and updated at source level for as long as it remains in use. Another technique used to plot a requested route is source-routing, as used in the dynamic source-routing protocol. AODV is one example of a reactive protocol.

– *Hybrid protocols.* The zone routing protocol is one example of a hybrid protocol, which attempts to combine the advantages of the two families of protocols discussed earlier. Nodes take a proactive approach to routing within their immediate vicinity, up to a certain distance (for example, three or four hops). If an application wants to send something to a node outside of this zone, a reactive search is triggered. Using this system, routes in the vicinity of a node are available immediately, and when the search needs to be widened, it is optimized: when a node receives a route-search packet, it knows immediately if the destination is within its own neighborhood; in this case, it can respond, or if not it propagates the request in an optimized manner outside of its proactive zone. Depending on the type of traffic and the routes requested, hybrid protocols of this kind can nevertheless combine the disadvantages of proactive and reactive routing methods: regular exchange of monitoring packets and inundation of the whole network when seeking a route to a distant node.

1.5. Conclusion

MANET networks have considerable potential for future wireless applications, thanks notably to their independence and autoconfiguration, low set-up costs (as no infrastructure is needed), adaptability to topological change, and the mobility of nodes. Nevertheless, these characteristics themselves present numerous problems and technical challenges. We would, therefore, like to consider the repercussions of these unusual characteristics on routing functions in MANET networks, notably in the absence of routers in the traditional sense, centralized entities responsible for routing in wired networks. These considerations will form the subject of the next chapter.

Chapter 2

Routing in MANETs

2.1. Introduction

Within a network, several routes usually exist for the transmission of a packet from a source to a destination node. The aim of a routing protocol is to find one or more optimal routes from a list of possibilities, minimizing the cost function. Mobile ad hoc networks (MANETs), as we saw in Chapter 1, have particular characteristics that distinguish them from wired networks, including the absence of a routing infrastructure, the unreliability of links, and a highly dynamic topology.

We must begin by considering whether existing Internet routing protocols are adapted to use in MANETs. We shall begin by reviewing routing protocols used in networks with a routing infrastructure and study the possibility of adapting these protocols to the demands of MANET. Next, we review the best-known routing protocols from current literature on the subject, with particular emphasis on the modifications needed to cope with the specific demands of ad hoc networks. Particular consideration will be given to the ad hoc on-demand distance-vector (AODV) [PER 03] and optimized link-state routing (OLSR) [CLA 03] protocols, the best candidates for normalization.

2.2. Internet routing protocols

Given a data packet, Internet protocol (IP) routing consists of determining the optimal route to take in order to reach the desired destination. In practice, a routing node (for example, a router) uses a routing protocol to generate a routing table in which all possible routes to a given destination are displayed.

There are two main phases in the operation of a routing protocol:

– A *neighborhood discovery* phase, where the routing node initializes and constructs its routing table. This phase occurs immediately after the node connects to the network, or in cases where the node has been restarted (after node failure).

– A *routing table update* phase that follows the discovery of topological changes in the network.

To determine the best route toward a destination, a routing node must have access to the topology of the network. It is tempting to suggest that nodes should exchange information on network topology, i.e. their routing tables. However, this solution is not foreseeable as an exchange of topological data of this type would occupy a considerable chunk of the available bandwidth to the detriment of the traffic of useful data.

Two possible solutions exist:

– Nodes exchange all topological data during the topological data update phase. This exchange would only be carried out between immediate neighbors. Routing protocols using this option are classified as *distance-vector routing protocols*.

– Nodes exchange topological data with all other accessible nodes in the network. Only part of the data (the part concerning topological changes) is sent during the topological data update phase. Routing protocols that operate in this way are classified as *link-state routing protocols*.

2.2.1. *Distance-vector routing protocols*

In routing protocols of this type, each node in the network sends its routing table to its neighbors. Data are sent either periodically upon expiration of a counter (i.e. a timeout) or when an update is triggered by topological modifications.

Upon receiving a neighboring node's routing table, a routing node uses the information to update its own routing table. For example, if a router *A* sends its routing table to a neighbor *B* and an entry in *A*'s table indicates that another node *X* is within two hops of *A*, *B* will update its table to show that node *X* can be reached by *B* via node *A* in two hops. Another routing node *C*, receiving the updated routing table of node *B*, will update its own table to show that it can reach node *X* via node *B* in three hops (unless *C* already has an available route to *X* in less than three hops), and so on and so forth.

Distance-vector routing protocols offer several advantages, including low bandwidth usage and low demands on the routing node in terms of processing power and memory capacity. However, protocols of this kind suffer from scalability and network convergence issues, particularly if the network develops loops (the *count-to-infinity problem* [SCH 02]). Effectively, if a node *A*, informs another node *B*, that it has a route to another node *X*, *B* may actually be part of the route between *A* and *X* without being aware of this fact. The route B→A→X would then not be optimal.

To give some examples of distance-vector routing protocols, the most used are routing information protocol [HED 88], IGRP [CIS 07], and AppleTalk RTMP [WAL 91].

2.2.2. *Link-state routing protocols*

Link-state routing protocols are more complex than distance-vector routing protocols.

Each routing node creates and updates a topological database in addition to the routing table. The topological database contains all topological information received from nodes in the network. This information is sent out in specific messages known as link-state advertisements.

Once the topological database has been updated, a routing node uses a shortest-path algorithm (typically Dijkstra's algorithm [WAN 07]) to calculate its routing table. To this end, it determines which are the other nodes in the network that are accessible to it and classifies them in a tree for which it is the root. This method allows the node to calculate the shortest path to any other accessible node in the network.

Link-state routing protocols are well suited to large and complex topologies. As routing nodes have a global vision of the network, the count-to-infinity problem does not arise.

However, link-state routing protocols require significant processing power and memory capacity at routing node level to store the topological database and to calculate routing tables from this database. In addition, the initial routing update is very costly in terms of bandwidth, as the topological databases of every node in the network must be sent.

The best-known link-state routing protocols are OSPF [COL 06] and IS-IS [PAR 06].

2.2.3. *Unsuitability of Internet routing protocols for MANETs*

In the previous paragraph, we have seen that routing requires an exchange of topological information between routing nodes, but this information transfer should not make excessive demands on the network in terms of resources. For this reason, it is necessary to limit the amount of information exchanged; in the case of link-state routing protocols, only part of the topological data is exchanged (that concerning topological changes). In the case of distance-vector routing protocols, all the topological information (i.e. routing tables) is exchanged, but only between immediately neighboring nodes.

In the case of MANETs, the need to minimize the consumption of resources by the routing protocol is even more strongly felt. First, there are no routing infrastructures (e.g. routers) in a MANET and mobile nodes themselves carry out routing. These nodes do not have the same processing power or memory capacity as a routing infrastructure, and their limited energy supply does not allow for a periodically recurring exchange of routing data.

Moreover, MANETs are highly dynamic due to the unpredictable, frequent, and sudden movement of nodes. This can create non-convergence problems within the network. This level of mobility can also generate "holes" in the network, for example, if a strategically important node moves out of the range of the other nodes.

In addition, the capacity and reliability of wireless links are limited in comparison with wired (fiber optic) links, due to potential perturbations in the radio channel, for example, interference and noise.

Finally, the fact that the radio channel is a shared resource creates security issues concerning the integrity of routing tables exchanged between different routing nodes.

Consequently, it would appear that Internet routing protocols are not suitable for use in MANETs under their current forms.

2.3. Classification of routing protocols in MANET

A number of routing protocols have been proposed for use in an ad hoc context, including adaptations of existing Internet routing protocols for use in ad hoc contexts as well as new protocols designed specifically for MANETs. We now attempt to classify these protocols that are best known from the existing literature on this subject.

2.3.1. *Table-driven routing protocols*

As existing Internet routing protocols perform well (especially link-state routing protocols, which are able to deal with complex topologies and large networks), it is worth trying to adapt them for use in MANETs.

Table-driven routing protocols use the same basic concept as Internet routing: at any given moment, each node can access a route to any destination accessible to the network. These routes are calculated based on topological information exchanged between nodes, then stored in a routing table which the node consults when deciding where to send a packet first in order to reach its destination.

Nevertheless, every table-driven routing protocol for MANETs makes one or more modifications to the Internet routing protocol model in order to adapt to an ad hoc context. In section 2.3.1.1, we present two examples of table-driven protocols which illustrate this adaptation.

2.3.1.1. *Destination-sequenced distance-vector routing (DSDV)*

DSDV [PER 94] is a table-driven distance-vector routing protocol for MANETs. Compared with link-state routing protocols, distance-vector protocols make fewer demands in terms of signalization traffic when exchanging topological data, as each node sends only this information to its immediate neighbors (see section 2.2.1).

However, as these protocols are based on the Bellman–Ford algorithm [ENC 07b], they have the drawback of being unable to deal with loops in the network. This makes them ill adapted for use in MANETs. The dynamic nature of MANETs, due to sudden, frequent, and unpredictable movement of nodes, makes it very likely that routing loops will appear in the network.

To overcome this problem, DSDV allocates a sequence number to each route in the table, the number being raised with each hop. This rule prevents routing loops. For example, if a node k is the next hop for k' when trying to reach a destination node i, then the route used by k' will have a sequence number $seq_{k'}$ higher than the sequence number seq_k of the same route in the routing table of node k:

$$seq_{k'} \succ seq_k$$

If a loop exists, k' will also precede k in the route toward i:

$$seq_{k'} \prec seq_k$$

This is evidently a contradiction. In this way, the problem of routing loops is resolved. DSDV has, however, the disadvantage of requiring regular updates to routing tables, which reduces the efficient bandwidth. Moreover, DSDV is not adapted to large networks. In the absence of specifications, there are no commercial implementations of this protocol.

2.3.1.2. *Optimized link-state routing protocol*

Where DSDV is an adaptation of distance-vector routing protocols for MANETs, OLSR is the equivalent adaptation of a link-state routing protocol. In contrast to distance-vector routing protocols, link-state routing protocols are not subject to routing loops and there are no problems with scalability. However, the exchange of topological data in a link-state routing protocol generates a large amount of traffic, an undesirable attribute in a MANET due to the limited resources available (see section 2.2.3).

The OLSR protocol [CLA 03] implements a new procedure for exchanging topological data within the network to reduce the volume of traffic involved. In OLSR, although all nodes are authorized to receive topological data messages, only a minimal number of nodes, known as multipoint relays (MPRs), are able to propagate these messages across the network. By definition, the MPRs of a given node are the minimum number of its immediate neighbors necessary to contact all its neighbors within two hops. This definition guarantees that topological data messages will be received by every node in the network.

2.3.1.2.1. Description of the protocol

RFC 3626 [CLA 03] defines OLSR as a core function, responsible for routing in the strictest sense of the term, and a group of auxiliary functions offering additional applications in specific scenarios, for example, if a node is responsible for maintaining connectivity between a MANET and another routing domain.

Core function

The core function of the OLSR protocol specifies the behavior of a node using OSLR as a routing protocol and of which the interfaces are configured with OLSR.

This core function is composed of the following subfunctions:

– definition of the format of packets and their transmission,

– detection of links,

– detection of neighboring nodes,

– selection and signalization of MPRs,

– diffusion of topological monitoring messages, and

– calculation of routes.

Auxiliary functions

In addition to the principal functions of OLSR, in certain situations other functions are desirable. For example,

– interoperability with non-OLSR routing domains in cases where a node is using multiple interfaces;

– propagation of redundant topological data; and

– advanced link detection.

2.3.1.2.2. Determination of multipoint networks (MPR)

OLSR uses packet flooding to transmit topological information within the network. Flooding, in its usual form, means that all nodes retransmit received packets. As we have seen, this mechanism can create loops.

In wired networks, one way of avoiding loops is not to retransmit any packet over the interface through which it arrived. In MANETs, the hopping technique used makes it essential that nodes retransmit packets in this way.

To deal with this issue, OLSR introduces the notion of MPRs. Under this system, a node will ignore a group of links and direct neighbors that serve no purpose in the calculation of shortest-path routes.

Thus, within the group of neighbors of a node, only a subgroup is considered relevant for routing. This subgroup is chosen in a way that makes the whole neighborhood accessible within two hops (i.e. neighbors of neighbors).

The nodes that make up this subgroup are known as MPRs. The MPRs of a given node *X* are a minimal group of immediate neighbors (i.e. one hop away) of *X* allowing it to access the whole neighborhood within two hops.

2.3.1.2.3. Selection of MPRs

In the classic flooding model, the transportation of a message through the whole network by retransmission is carried out using the following rule: "a node retransmits a message if, and only if, it has not already received it."

Diffusion by MPRs lessens the number of retransmissions by using the following rule [CLA 03]: "a node retransmits a message if, and only if:

– it has not already received it and

– it has just received the message from a node for which it acts as a MPR."

The flowchart in Figure 2.1 illustrates the steps involved in the selection of MPRs.

2.3.1.2.4. Calculation of the routing table

Each node maintains a routing table that enables it to direct traffic toward other nodes in the network. This table is filled in based on information contained in the information base on local links and the base of topological information.

Consequently, if change occurs within either of these two bases, the routing table is recalculated to update the information concerning routes toward different destinations.

An entry in the routing table of an OLSR node (Table 2.1) shows that the node identified as R_dest_addr is estimated at R_dist hops from the local node that the symmetric neighbor with the address R_next_addr is the next hop toward R_dest_addr and that this immediate neighbor is reached via the local interface R_iface_addr.

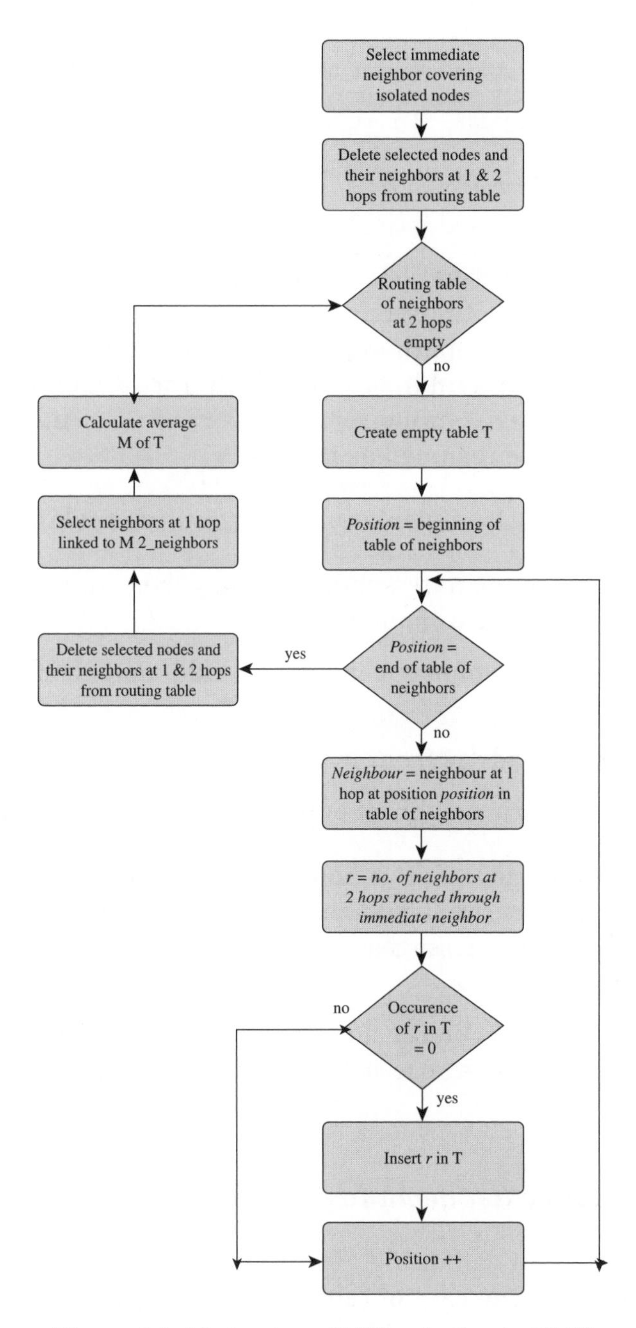

Figure 2.1. *Mechanism of MPR selection in OLSR*

R_dest_addr	R_next_addr	R_dist	R_iface_id
R_dest_addr	R_next_addr	R_dist	R_iface_id

Table 2.1. *Format of entries in the OLSR protocol routing table*

Only those destinations to which a route is known are included in the routing table. To produce the routing table of a node X, a shortest-path algorithm is applied to the network assimilated to a graph orientated in the direction X→Y, where Y is any symmetric neighbor of X. The routing algorithm can be resumed as follows:

ALGORITHM 2.1.– Calculation of the routing table in OLSR

1. Add all immediate symmetric neighbors to the routing table with a distance of one hop.

2. For each symmetric neighbor, add neighbors at two hops which are

   (a) not already present in the routing table and

   (b) symmetric neighbors of the node under consideration.

3. For every node at two hops' distance added to the table, add all neighbors (i.e. nodes three hops from the source node) known from TC messages and not already present in the routing table.

4. Repeat Step 3 until an iteration n occurs for which no more entries at distance $n + 1$ can be found in the routing table.

2.3.2. *Reactive (on demand) routing protocols*

Rather than adapting Internet routing protocols to MANETs as in the case of table-driven protocols, on-demand routing protocols introduce new routing concepts specific to the ad hoc context.

The basic principle of reactive protocols is to seek a route to a destination only in case of need, hence the alternative denomination of "on-demand routing." At a given moment, a node *S* may not possess a route to a destination node *D*. If this is the case and *S* wishes to send traffic to *D*, then *S* launches a path-seeking procedure to find a route to *D*. Usually, *S* sends a request to other nodes in the network to see if any node possesses a route to *D*. Any node in possession of such a route will reply. *S* will then transmit traffic to one of these nodes, which then forwards it on to *D*.

Generally, a reactive routing protocol has the following two basic mechanisms:

– A route discovery mechanism, by which a node may seek a route to a destination.

– A route maintenance mechanism, through which a node maintains its routes to a destination.

This concept will be illustrated using the two best-known examples of reactive routing protocols.

2.3.2.1. *Dynamic source-routing (DSR)*

DSR is a reactive (on demand) routing protocol denoted by the experimental RFC 4728 of the IETF [JOH 03]. When a node *S* wishes to send traffic to a destination *D* for which it does not possess a route, *S* starts a procedure of route discovery. To this effect, a route request (RREQ) message is sent to all its neighbors. If *D* is in fact a neighbor of *S*, this neighbor sends a route reply (RREP) message to *S*. Otherwise, the neighbor propagates the RREQ message to its own neighbors. This procedure is repeated until the RREQ message reaches the destination node *D*, or until the RREQ message has passed through a number of nodes higher than the tolerated maximum, *Time to Live* (TTL, set at anywhere from 1 to 255 hops).

Assuming node *D* receives the RREQ message within TTL hops, *D* sends an RREP message to the source node *S* along the same route taken by the RREQ message. The RREP message indicates to node *S* all the intermediary nodes on the route to *D*. *S* updates its cache routing table by adding this route, then sends its traffic to *D*.

If the position of *D* in the network topology changes, the route becomes invalid. Such changes are detected when an intermediary node between *S* and *D* notices that the next hop on the route specified by the packet is no longer attainable. The node then states the route maintenance procedure by sending a route error (RRER) message, indicating to node *S* that the route to *D* is no longer valid. Consequently, node *S* deletes the route in question from its cache routing table and begins a new route discovery operation to find a new means of reaching *D*.

Nevertheless, DSR has the disadvantage of saving the IP address of every node involved in a route within the packet being transmitted. This considerably increases the size of the packets' headers, particularly in cases where IPv6 addresses are used.

2.3.2.2. *Ad hoc on-demand distance-vector*

AODV is a reactive routing protocol specified in the RFC 3561 of the IETF [PER 03]. It functions in a very similar way to DSR, while considerably reducing the time taken by DSR to find a route.

In AODV, an RREQ message does not need to reach the desired destination for an RREP message to be sent in reply. All that is needed is for the RREQ message to reach an intermediary node that already has a route to the destination in its routing table; this intermediary then sends an RREP message back to the source. Subsequently, the source chooses the shortest possible route from among the RREP messages received (the unit of measurement used in this protocol being the number of hops).

2.3.2.2.1. Description of the protocol

AODV is a reactive (on-demand) routing protocol for use in MANETs. It allows mobile nodes to obtain routes on demand and does not necessitate the maintenance of routes to destinations, a fact that considerably reduces the size of the routing tables saved by each node and the monitoring traffic linked to these tables.

Effectively, as long as the extremities of a connection are linked by a valid route, AODV does not participate in routing. When a route is required, AODV uses a route discovery mechanism to connect the pair of nodes in question. In this case, the source node diffuses a RREQ message in broadcast mode to ask other nodes in the network if they have a route to the required destination. The reception of this message by intermediary nodes allows the latter, on the one hand, to dispose of a "fresh" route to the source in question, and on the other to participate in routing within the network by sending RREP messages back to the source if a route to the destination is known.

Moreover, like every on-demand routing protocol, AODV possesses a route maintenance mechanism, which notifies other nodes if a link is cut and tries to "repair" local links.

Losses of connectivity are announced to the network by the diffusion of RERR messages. For this system to work, each node must maintain a list of its precursors in a given route. When a link is cut, the node creates a list of other nodes likely to be affected, then advises them of the event using RERR messages as described earlier.

Nevertheless, a node may attempt to repair the link if the other extremity is within its neighborhood. In this case, it makes a series of retransmissions of RREQ messages to check whether the connection has been lost. If the connection is re-established, the node informs the other nodes that the route is still valid.

Finally, AODV – like all other routing protocols – maintains a routing table containing both routes to all immediate neighbors to which a symmetric link is detected and routes to sources for which RREQ messages have recently been received.

2.3.2.2.2. Route discovery mechanism

A route discovery mechanism is the process that enables routes to destinations to be found when requested by sources. In this section, we shall examine this procedure in detail and the role each node in the network – source, intermediary nodes, and destination – plays in the process.

Completing the routing table and the creation of precursor lists

When a node receives an AODV control packet from a neighbor, or when it creates or updates a route to a particular node or subnetwork, it consults its table to see if a corresponding entry exists. If the search is unsuccessful, the node creates a new entry in the table. Otherwise, updates are only made after the sequence numbers of the destination of the new entry and the destination of the old entry have been compared.

Finally, for every route saved in the routing table, the node maintains a list of precursors that can transport traffic on that route. The precursor list contains a list of neighboring nodes, which have transmitted or generated a RREP message.

RREQ message generation

A node generates an RREQ message when it wishes to transmit data to a destination not already present in its routing table. The node begins by saving the ID of the RREQ message that it will send and the corresponding instant of generation (PATH_DISCOVERY_TIME) so as not to have to process the message when it is propagated back by neighbors.

Next, the source sends out the RREQ message and starts a timer while waiting for a response. If no response has been received by the time the counter expires, the source may retransmit the RREQ message after a certain time (known as backoff time) has elapsed. This can continue until a maximum number of RREQ_RETRIES attempts is reached.

Control of RREQ message dissemination

To limit the scale of the broadcast, a mechanism is needed to prevent unnecessary dissemination of RREQ messages in the network. This mechanism is the *expanding ring search technique*.

Using this method, the source begins its search with a TTL equal to TTL_START and defines a time, RING_TRAVERSAL_TIME, at the end of which it expects to receive an RREP response. If this time passes and the source has still not received a reply, TTL is increased by one step, TTL_INCREMENT. This is repeated as many times as necessary until TTL reaches a threshold TTL_THRESHOLD at which the value of TTL is equal to the diameter of the network (NET_DIAMETER). By increasing the TTL each time, the radius of the search zone is increased, effectively making expanding rings, hence the name of the technique.

It should be noted, however, that this progressive pattern of searches can be avoided by setting TTL_START at the diameter of the network from the beginning.

Processing and transportation of RREQ messages

When a node receives an RREQ message, it first updates its route toward the previous hop (i.e. the node it received the message from). Next, it checks to see if the same message has already been received; if this is the case, it destroys the packet "silently" (silent discard), i.e. without announcing this operation to the other nodes. If the request has not previously been received, the node raises the number of hops in the packet then seeks a route back to the source in its table. Depending on the result of this search, the node can update or insert a route to the source.

This being the case, the node then looks at the destination address in the packet to carry out the actual routing. To this end, it checks that it is not itself the destination or if it has a route to the destination in its own routing table; in these cases, an RREP message is generated and sent back to the source.

Generation of RREP messages

A node generates an RREP message in two cases:

– if it is the destination or

– if its routing table contains a route to the destination for which the sequence number is greater than or equal to the sequence number indicated in the RREP message.

During the generation of an RREP message, the node inserts the destination address and the sequence number contained in the RREQ message received. The RREP is then sent back to the source along the same route by which it arrived. If the node sending the RREP message is not the destination but an intermediary node, it also gives the distance that separates it from the destination.

Reception and transportation of RREP messages

Upon receiving an RREP message to send back to the source, a node seeks a route to this source in its routing table. Then, it raises the number of hops in the RREP packet. Subsequently, the node consults its routing table and updates its route to the destination requested in the original RREQ message. The update process is similar to that discussed in relation to the treatment of RREQ messages. The node also updates its list of precursors for the route in question. Finally, the node transmits the RREP message to the next hop on the path to its destination.

2.3.2.2.3. Route maintenance mechanism

The route maintenance mechanism is the second defining mechanism of on-demand routing protocols. It serves to identify actions to be taken by nodes to guarantee local connectivity and to propagate messages announcing the rupture of a link to other nodes. In this section, we shall present the operations effectuated in the course of this procedure in detail.

Maintenance of local connectivity

RFC 3651 [PER 03] demands that each node keeps track of its connectivity to neighboring nodes and the connectivity of other nodes to itself.

Connectivity maintenance notifications can be sent either via the notification mechanisms of the link layer, or via the intermediary of connectivity tests using ICMP echo request messages or RREQ messages sent in unicast mode to the neighboring node only.

If a link cannot be detected using any of the methods already described, the node assumes that the link is lost and takes one of the actions described in the following sections.

Processing of RRER messages

A node generates an RRER message in any of the following cases:

– if a link to the next hop in an active route within the routing table is found to be cut at the moment of data transmission;

– if a data packet is received which is destined for a node toward which it has no active route in the routing table; and

– if an RERR message is received from a neighbor.

Generally speaking, the following steps are involved in the creation and transmission of RRER messages:

– identification of the routes in question as invalid;

– listing the destinations concerned;

– identification of neighbors (present in the precursor list for the route in question) affected by the rupture of the link; and

– sending the appropriate RERR messages to these neighbors.

Depending on the number of precursors involved, RERR messages can be sent in broadcast mode (for several precursors), unicast mode (for a single precursor), or by repeated unicast to each precursor in turn if broadcast mode is not appropriate.

Local repair

When a link in an active route is cut, a node may choose to repair it if the distance between it and the other extremity of the broken list is not greater than a fixed maximum number of hops, MAX_REPAIR_TTL. To repair the link, the node raises the sequence number of the destination then diffuses an RREQ message toward it.

If no RREP response is received from the destination at the end of the route rediscovery phase, the repair process is considered to have failed and the node begins to generate RERR messages as described earlier.

If, on the other hand, the node receives one or more RREP messages for the diffused request, it generates an RERR message to indicate to the other nodes that the route containing the defective link should not be deleted, then updates its routing table, replacing the old broken route with a new one.

2.3.3. *Hybrid routing protocols*

Certain so-called hybrid protocols use both proactive and on-demand routing concepts with the aim of combining the advantages of both types of protocol. One protocol of this type is zone routing protocol (ZRP) [HAA 02].

ZRP divides the network into interlinked routing zones. Within each routing zone, an intrazone routing protocol (IARP) operates proactively to supply all nodes in the zone with routes to the other nodes in the same zone. In this way, all nodes in the same zone can access the topology of the whole zone. Moreover, the exchange of topological data involved does not have a noticeable effect on the available bandwidth, given that this information applies only to a small network.

In addition to these proactive aspects, ZRP uses a reactive interzone routing protocol to find routes between nodes in different routing zones. The route-seeking procedure uses RREQ messages, which are propagated across the whole network.

It should be noted that the size of routing zones is a deciding factor in obtaining optimum functionality using this protocol.

2.3.4. *Hierarchical routing protocols*

Hierarchical routing protocols establish hierarchies among nodes in the network. To give an example, cluster-based routing protocol [JIA 99] divides the network into clusters, defining one node as the head of each cluster. When a source node seeks a route to a destination, RREQ messages are sent only to the heads of clusters within the network. Each head of cluster checks if the target destination is within its cluster, in which case it forwards the RREQ message to the node in question. The destination will then send an RREP message back to the source and a route is established between the two nodes.

Although this approach to routing reduces the flooding of the network by RREQ messages, it increases the load at the level of the heads of cluster, depleting their resources rapidly. This approach raises questions on the optimum manner of cluster formation and choice of heads of cluster, which are generally whole-NP problems [ZHA 05].

2.3.5. *Geographic routing protocols*

Geographic routing is based on the idea of sending traffic to the geographic location of a node rather than its IP address. This can be interesting insofar as a route to a destination is no longer necessary: traffic can simply be directed toward the geographic area of the target. Several geographic routing protocols have been proposed, including ALARM [BOL 04], DREAM [BAS 98], LAR [KO 98], and SiFT [CAP 05] among others.

As an example, location-aided routing (LAR) [KO 98] proposes the use of a geographic localization system such as global positioning system to localize nodes. Each source uses this localization information to calculate a zone where it expects to find the destination at a given moment (the expected zone). Subsequently, a requested zone is calculated, including the expected zone, to establish a group of nodes to which an RREQ message will be sent in order to find the exact location of the destination.

Distance routing effect algorithm for mobility (DREAM) [BAS 98] takes as its departure point the statement that the further two nodes are from each other, the longer it takes them to notice the mobility of the other. Because of this, DREAM proposes to regulate the frequency of localization updates in routing tables in relation to the distance between nodes.

2.3.6. *Routing protocols with power control*

Energy constraints are an important feature of MANETs as nodes generally use limited-capacity energy sources, usually batteries.

As data transmission is a determining factor in the energy consumption of a node, certain routing protocols take energy constraints into consideration in routing.

These routing protocols follow the basic principle that the energy required to transmit a signal is proportional to the distance which separates the sender from the receiver by a factor α, greater than 2, which depends on the nature of the surroundings [ISL 07].

In optimum conditions where $\alpha = 2$, to transmit the signal to a point halfway between sender and receiver would take one-fourth of the energy required for direct transmission. If another node halfway along the route from source to destination is prepared to sacrifice another one-fourth of the total energy required for transmission from source to destination, then the total energy consumed by the transmission would be one-half of what would be required if the sender had transmitted the signal to the receiver directly.

Evidently, although this approach reduces energy consumption, it also increases the time required for transmission by adding to the number of hops in the route, potentially creating supplementary delays in case of queues and retransmissions.

PARO [GOM 01] and EADSR [BRO 03] are two examples of routing protocols with power control.

2.3.7. *Multicast routing protocols*

Multicast allows the same content to be sent to several destinations simultaneously. This mode of transmission has a number of useful applications, for example in videoconferencing, advertising, and television via IP. Certain protocols aim to offer multicast routing adapted to MANETs.

To give an example, ad hoc QoS multicast (AQM) [BUR 05] is a multicast routing protocol with quality of service (QoS). AQM determines the bandwidth available to each node based on the resources made available to the neighborhood as a whole by other nodes. When a new AQM node wants to subscribe to a multicast session, AQM checks whether the available resources are sufficient to cater for this node. If enough resources are available, they are reserved for the use of the new node.

Multicast ad hoc on-demand distance-vector [ROY 99] is a multicast extension of the AODV protocol [PER 03] presented above (see section 2.3.2.2). RREQ messages are sent to identify multicast routes, i.e. routes to several destinations. Afterwards, multicast RREPs travel through member nodes of the multicast group to arrive back at the source.

2.4. Conclusion

The constraints of MANETs have rendered the task of routing particularly difficult. In this chapter, we have presented various approaches to Internet routing and demonstrated their unsuitability for use in MANETs. We have also given a classification of routing protocols proposed for use in ad hoc networks. Nevertheless, despite the variety of solutions available, at the time of writing none has yet been normalized for routing in MANETs.

Chapter 3

Performance Evaluation of OLSR and AODV Protocols

3.1. Introduction

The choice of routing protocol in an ad hoc network is not always obvious, as each protocol is adapted to very specific environments. Various comparisons have been made between the principal reactive and proactive protocols based on different performance criteria, such as packet delivery rate and the time taken for a packet to travel from source to destination. In this chapter, we first deal with the operating mechanisms of the ad hoc on-demand distance-vector (AODV) and optimized link-state routing (OLSR) protocols. We then present a comparative study based on the results of simulations, which allow us to reach a conclusion regarding the relative performances of the two protocols.

The choice of the AODV and OLSR protocols as subjects for study in the course of this chapter owes nothing to chance; above all, this choice stems from a need to compare an established protocol (AODV) with a more recent one (OLSR).

These protocols are very different from each other, given they belong to two different families. The AODV protocol is an on-demand (reactive) protocol with a flooding-based method of route discovery. OLSR, on the other hand, is proactive (table-driven), a descendent of distance-vector and link-state protocols; OLSR therefore maintains routing tables with information on the state of the network which are updated periodically.

These two protocols are therefore completely opposite in terms of basic operational principles; more details will be given later.

Another reason for the choice of OLSR and AODV for this comparison is the size and importance of studies undertaken by researchers in this domain since the emergence of MANETs. As we stated above, MANETs are currently undergoing standardization.

It should be noted that the request for comments (RFC) of the AODV protocol is 37 pages long, whereas the OLSR is 77 pages long, that is more than twice the length of AODV. Thus, both have their shortcomings and could be improved; hence, the necessity of studying the contribution of OLSR in relation to the competition with the aim of improving OLSR and the quality of service (QoS) provided.

To evaluate the performance of OLSR and AODV, it is first essential to study the ways in which the two protocols work.

3.2. The AODV protocol

The AODV protocol is a reactive routing protocol, designed essentially for use in ad hoc networks. AODV is one of the best-known routing protocols and has attracted a great deal of interest from researchers and the scientific community. It appears on the list of protocols to be standardized in the near future. AODV fits into the context of "best effort" routing. It is reactive in nature and based on two central mechanisms that serve to establish and maintain active routes.

The AODV protocol seeks routes when a source asks to send packets to a destination. Once established, the route between the source and the destination must be maintained for the duration of the communication. The establishment and maintenance of routes are carried out by exchanging three types of messages:

– route request (RREQ),

– route reply (RREP), and

– route error (RERR).

These messages are received via the user datagram protocol (UDP) protocol using an Internet protocol (IP) header. Thus, the node requesting to send data uses its IP address as a source of IP address. For diffusion messages, the IP address (255.255.255.255) is used. The encapsulation of data follows the pattern set out in Table 3.1.

MAC header	IP header	UDP header	AODV header	Data

Table 3.1. *Encapsulation of data in a network using AODV*

3.2.1. *Route establishment*

3.2.1.1. *Path discovery*

The path discovery process is launched when a source node wishes to communicate with a destination node for which it has no routing information in its table. Each node maintains two separate counters: a node sequence number (a number generated by the node itself to guarantee information is up-to-date) and a broadcast_id (the unique identifier of a packet sent out by the source). The source node starts the path discovery process by sending out a RREQ packet to its neighbors.

The pairing of <source address, broadcast_id> gives the unique identity for an RREQ. Broadcast_id is raised each time the source generates a new RREQ. Each neighbor responds to the RREQ either by sending a RREP to the source if it is the destination node, or by diffusing the RREQ packet to its own neighbors (if it has a valid path for the requested destination) after raising the hop_count (a counter which tracks the number of hops).

It should be noted that a node can receive multiple copies of the same packet from different neighbors. When an intermediary node receives an RREQ with the same broadcast_id and source address more than once, it rejects the superfluous RREQs without rediffusing them, thus avoiding overloading the network.

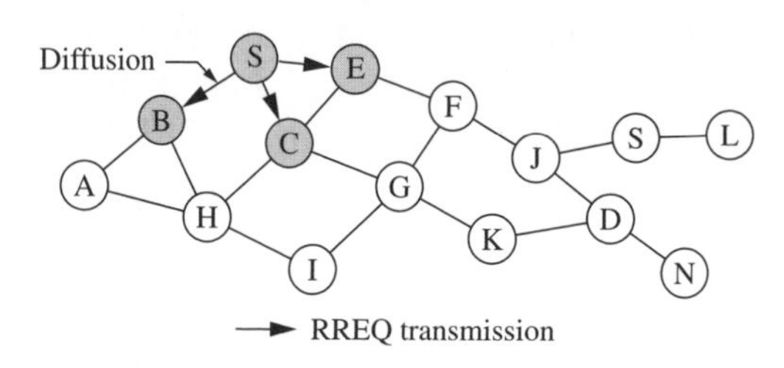

Figure 3.1. *Diffusion of RREQ messages in a network using AODV*

To further limit the load on the network, AODV uses a progressive method to extend route searches. The request is initially diffused within a fixed number of hops. If the source does not receive a response within a set time period, another message is sent out over a wider area (i.e. the maximum number of hops is increased). If there is still no reply, this process is repeated up to a fixed maximum number of tries, after which the destination is declared inaccessible.

3.2.1.2. *Reverse path setup*

As an RREQ moves from a source to different destinations, it automatically saves the reverse path back to the source.

To save the reverse path, a node saves the address of the neighbor from which it received the first copy of the RREQ. These reverse path entries are maintained long enough for the RREQ to cross the network and for the transmitter to receive a response.

Two sequence numbers are included in the RREQ: the sequence number of the source and the sequence number of the destination known to the source. The source sequence number is used to maintain the freshness of information on the route back to the source. The destination sequence number indicates how fresh the path to the destination must be to be accepted by the source. Sequence numbers for a different route must be greater than or equal to those included in the message.

3.2.1.3. *Forward path setup*

Finally, an RREQ arrives at a node, which is either the destination itself or possesses a route for the destination.

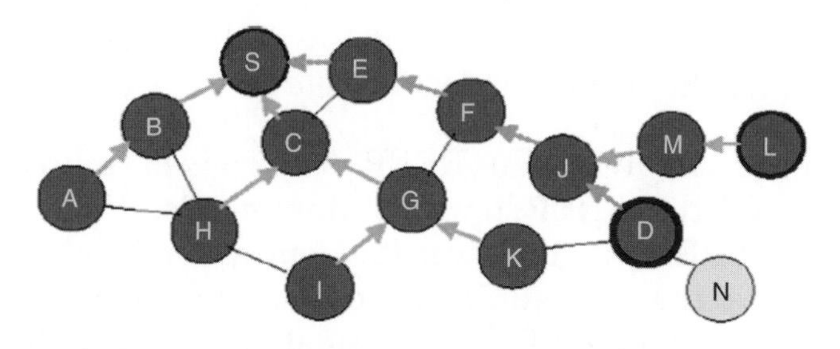

Figure 3.2. *Sending of RREP message to confirm a path*

The receiving node first checks that the RREQ message was received over a bidirectional link. If an intermediary node has a route entry for the required destination, it checks whether the route is current by comparing its sequence number to that of the destination in the RREQ message. If the sequence number contained in the RREQ for the destination is higher than that saved by the intermediary node, then the node should not use its saved route to reply to the RREQ and updates its sequence number for the destination in question. The intermediary node then rediffuses the RREQ.

In other words, the intermediary node may only respond if, in its routing table, it has a route for which the sequence number is greater than or equal to that contained in the RREQ. If not, it updates its routing table or creates a new route entry in the table. If the node has a current route to the destination and if the RREQ has not already been received, the node sends an RREP message by unicast to the neighbor from which it received the RREQ.

When a diffusion packet reaches a node that can supply a route to the destination, a return path back to the source of the RREQ has already been established. When the RREP is transmitted to the source, each node on the route saves a pointer in the exact direction on the node from which the RREP arrived, updates its routing table entries for the source and the destination and saves the latest sequence number for the requested destination.

A node, receiving an RREP for a given source node, propagates the RREP towards this source. If it receives another RREP, it updates its routing information and propagates this second RREP exclusively in cases where the sequence number for the destination is higher than that contained in the first RREP, or if the sequence number is the same but the hop_count is lower. All other RREP messages will be deleted. This reduces the number of RREP messages being propagates towards the source while keeping the most up-to-date routing information and making routing as rapid as possible.

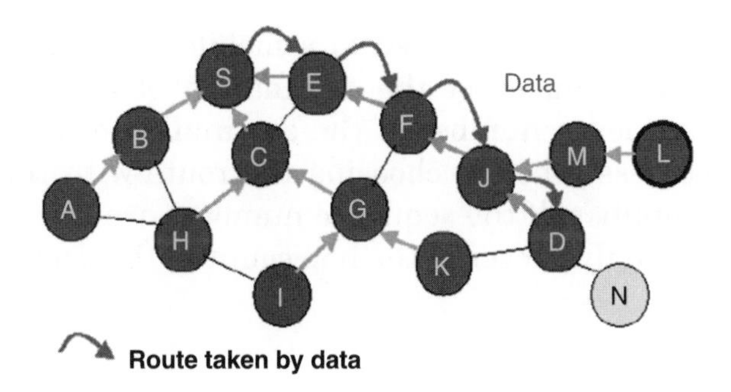

Figure 3.3. *Transmission of data by the AODV protocol*

The source node can begin transmitting data as soon as the first RREP is received and can update its routing information later if a better route is discovered subsequently.

3.2.1.4. *Routing table management*

In every routing table entry, the addresses of active precursors through which packets traveling to the desired destination were received are also saved. A neighbor is considered as active if it has generated or relayed a packet for the destination in question during the most recent period on the active_route_timeout counter. This information is stored so that all active source nodes can be informed when a link is broken in the path to the destination. A route entry is considered to be active if it is in use by any active neighbor. The route between a source and a destination, followed by packets via active route entries, is known as an active path.

It should be noted that, as in DSDV, all routes in the routing table are labeled with the sequence numbers of the destination, which prevents the formation of routing loops even in extreme conditions such as when nodes are extremely mobile. Each node maintains a routing table for each destination of interest.

If a new route is proposed to a mobile node, it compares the sequence number of the destination of the new route with the sequence number of the destination of the route it already possesses before choosing the route with the higher sequence number. If the sequence numbers are the same, the new route is only chosen if the hop count to the destination is lower.

3.2.2. *Path maintenance*

The movement of nodes not belonging to an active path does not affect routing towards the destination of this path. If the source node moves in the course of an active session, it may reinitialize the route discovery process in order to establish a new route to the destination.

When the destination or intermediary nodes move, a special RREP message is sent to affected source nodes. Periodic HELLO messages can be used to confirm the presence of symmetric links and to detect link ruptures.

Alternatively, link ruptures can be detected using link_layer acknowledgement messages. This method is considerably quicker.

A link rupture can also be detected if attempts to send a packet to the next hop fail. Once the next hop becomes unreachable, the previous node on the path propagates an unsolicited RREQ message to all its active neighbors further back along the route. This RREQ contains a fresh sequence number (i.e. the previous sequence number is raised) and a hop_count of infinity. Nodes receiving this message send it to their own active neighbors, and so on and so forth.

This process continues until all active source nodes have been advised. The process will end at a fixed moment, as AODV will only maintain routes without loops and an ad hoc network contains a finite number of nodes. At the end of the process, the source nodes may restart a route discovery process if they still require an open route to the destination concerned.

To find out whether a route is still necessary, a node may check if the route has been used recently and check the control sequences of protocols at higher levels to see if connections using the destination in question are still open. If the source node (or any other node on the preceding route) decides to rebuild the route to the destination, it sends out an RREQ with a destination sequence number equal to the preceding incremented sequence number to inform other nodes that is constructing a new route, and that a node should not respond if it still considers the previous route to be active.

Nodes recognize their neighbors using two different methods. When a node receives a broadcast from a neighbor, it updates its local connectivity information to ensure that this information contains details on the neighbor in question.

In cases where a node has not sent any packets to its active neighbors before the hello_interval has elapsed, it sends a HELLO message (a nonsolicited RREP) containing its identity and sequence number. The sequence number of the node remains the same for the transmission of HELLO messages. This message cannot be diffused outside the node's immediate neighborhood as it has a time to live (TTL) of one. On receiving the packet, neighbors of the source node update their local connectivity information concerning this node.

The reception of a HELLO message from a new neighbor or failure to receive consecutive HELLO messages from a node, which was previously part of the neighborhood indicates changes in local connectivity.

Failure to receive HELLO messages from inactive neighbors does not spark any reaction at protocol level. If HELLO messages are not received from the next hop on an active path, a link rupture notification is sent to active neighbors using this next hop. This notification is carried out using RERR messages.

3.3. The OLSR protocol

The OLSR protocol in its current form was proposed by the Institut National de Recherche en Informatique et en Automatique (INRIA) in the course of the HIPERCOM project. The OLSR protocol belongs to the link-state family of protocols. The feature that sets OLSR apart from other protocols of the same type is its use of partial topology.

In other words, nodes do not diffuse information on all the links they have with their neighbors, but only a subgroup of links which allow them to contact these neighbors. This reduces the size of monitoring packets diffused within the network.

The OLSR protocol is based on the multipoint relay (MPR) technique. This technique optimizes the diffusion of monitoring messages in the network and reduces bandwidth consumption. The MPR technique causes less disturbance in the network than diffusion by flooding.

OLSR is a proactive (table-driven) protocol. To maintain an up-to-date image of network topology, OLSR nodes periodically exchange control packets. Information extracted from these packets is saved for a limited period, known as information validity time.

In this way, the protocol is able to cope with occasional losses of control packets. This kind of loss is entirely natural as the radio environment is very vulnerable to transmission problems, such as interference, mobility, and obstacles. However, this does not affect the operation of the network as packets are given sequence numbers. Packets with a sequence number older than that of packets already received will simply be rejected.

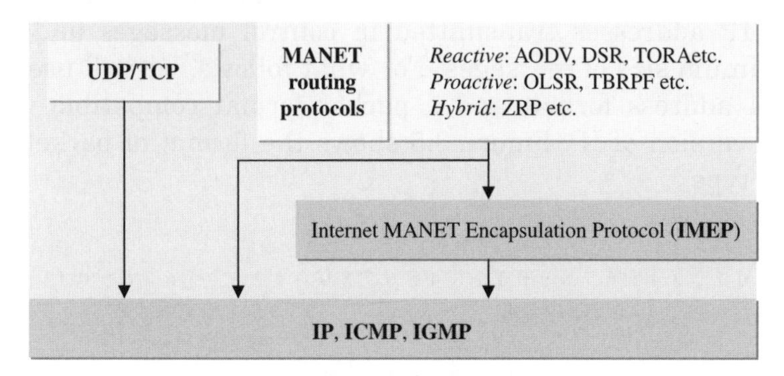

Figure 3.4. *Hierarchy of MANET routing protocols*

Information extracted from these control packets allows various tables to be updated as well as the transmission of data through routing tables. Like all routing protocols (Figure 3.4), OLSR does not deal directly with data packets. The IP layer is responsible for processing data packets and pathways using information contained in the routing table.

OLSR uses a standard format for IP packets, meaning that there is no need to modify IP stacks. The OLSR protocol uses a unified packet format for control messages. These packets are sent in UDP datagrams. Each packet encapsulates one or more messages at a time, optimizing access to the radio channel.

MAC header	IP header	UDP header	OLSR header	Data

Table 3.2. *Encapsulation of data in a network using OLSR*

3.3.1. *Format of OLSR packets and node addresses*

OLSR uses IP addresses as the unique identifier of nodes in the network. It can function over multiple communication interfaces. For this reason, each network node chooses the address of one of its interfaces as a main address.

OLSR can be used with both versions of IP, version 4 (IPv4) and version 6 (IPv6). The difference lies in the size of the IP addresses transmitted in control messages and the minimum size of messages. For what follows, we will use the IPv4 address format and a packet format compatible with this version of IP. Figure 3.5 shows the format of packets of this type.

0	1	2	3
0 1 2 3 4 5 6 7 8 9 0 1 2 3 4 5 6 7 8 9 0 1 2 3 4 5 6 7 8 9 0 1			
Packet Length		Packet SequenceNumber	
Message Type	Vtime	Message Size	
Originator Address			
Time To Live	Hop Count	Message Sequence Number	
Message			
Message Type	Vtime Message	Size	
Originator Address			
Time To Live	Hop Count	Message Sequence Number	
Message			

Figure 3.5. *Format of an OLSR packet*

The OLSR packet has several component parts:

– *Packet length*. Size of the packet in bytes.

– *Packet sequence number*. Each packet transmitted has a sequence number.

– *Message type*. One Hello, two topology control (TC), three multiple interface declaration (MID), and four host and network association (HNA).

– *Vtime*. Validity time, i.e. the length of time after reception of a message for which the receiving node should consider the information in the packet as valid, assuming a more recent update is not received in the meantime. Validity time is calculated as follows:

$$C\left(1+\frac{a}{16}\right)2^{b} \text{ (in seconds)} \qquad [3.1]$$

where a is the integer represented by the four most significant bits (MSB), b is the integer represented by the four least significant bits (lsb), and C is a multiplication factor (the value of C depends on the application).

– *Message size*. The size in bytes, from the start of *message type* to the beginning of the following *message type* or the end of the packet if there is no following message.

– *Originator address*. It contains the main address of the node that generated the message.

– *TTL*: 0–255. It contains the maximum number of hops the message can make. Before a message is retransmitted, the TTL should be reduced by 1. When a node receives a message with a TTL of 1, the message is not retransmitted.

– *Hop count*. It contains the number of hops a message has made. Before a message is retransmitted, the hop count is increased by 1. The initial value of hop count when the message leaves the source is 0.

– *Message sequence number*. The source node assigns a unique identifier to each message in the form of a message sequence number. Each time a node generates a message, it increases the sequence number by 1. These sequence numbers are used to ensure that messages are not retransmitted more than once per node.

– *Data*: It contains the actual message, whether HELLO, TC, MID, HNA, or a data packet.

3.3.2. Operation of the protocol

The OLSR protocol contains two principal mechanisms, one for neighborhood detection and one for topology management. For this, four types of control messages are used: HELLO, TC, MID, and HNA.

Neighborhood sensing is carried out using HELLO packets. The dissemination of topological information is carried out by the diffusion of TC packets using optimized diffusion and MPRs. TC messages contain a list of links in the neighborhood of the node managing the packet.

The OLSR protocol also takes into account all interfaces linked to a mobile unit using MID messages. Thus, nodes can make use of all available routes independent of the type of interface used at each hop. The OLSR node chooses one of its interface addresses as a main address, which it then uses as a reference in control messages.

HNA messages are used to declare subnetworks and hosts (outside of MANET) reachable by a node acting as a gateway. In what follows, we shall concentrate on the first three types of control messages.

3.3.2.1. *Neighborhood sensing*

As a derivative of classic link-state protocols, OLSR maintains a variety of information tables. These are updated every time control messages are received and information is stored with every message sent. Nodes store a variety of different tables in cache:

– *MPR selector Set*. This table contains all the nodes the local node has selected as MPRs.

– *Neighbor set*. All neighbors at one hop are saved in this table. It is updated dynamically through link set data. Information on symmetric and asymmetric link neighbors is stored in this table.

– *Two-hop neighbor set*. This table contains information on nodes accessible via one-hop paths, including the node in question itself. Note that this table may contain information on the same nodes appearing in the neighbor set table.

HELLO messages have three different roles in the OLSR protocol. HELLO messages are sent to neighbors at one hop (for link sensing and neighbor sensing), to neighbors at two hops (for two-hop sensing), and to declare MPRs (MPR selector sensing).

3.3.2.1.1. Link-state sensing

The OLSR protocol uses HELLO messages for link-state sensing. A simple example of neighbor sensing is given in Figure 3.6. *A* sends an empty HELLO. On receiving this message, *B* saves *A* as an asymmetric neighbor as it does not find its own address in the HELLO message. After 2 s, *B* itself sends a HELLO message including the address of *A*. Upon receiving the HELLO sent by *B*, *A* will find its own address in the message, and so declare *B* as a symmetric link. *A* subsequently includes the address of *B* in a HELLO message and *B* will save *A* as a symmetric neighbor on receiving this latest message.

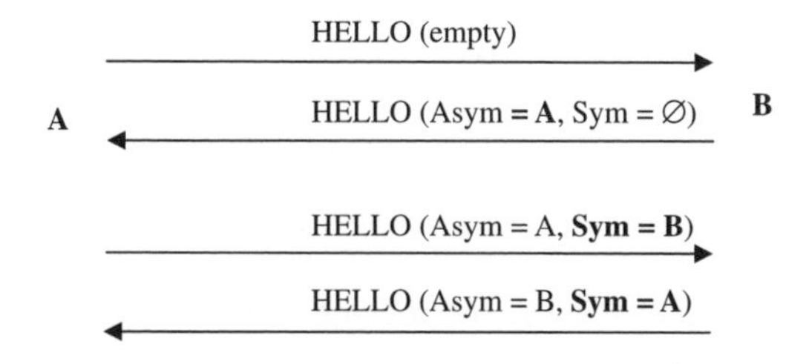

Figure 3.6. *Neighbor sensing by exchange of HELLO messages*

3.3.2.1.2. Direct neighbor sensing

Neighbor sensing is used to populate the table of direct neighbors using the principal addresses of nodes. Each time a new entry is detected during a link-state change, the table is updated. A neighbor is declared to be symmetric if at least one bidirectional link is present in the link table allowing a local interface to connect to neighbors. If a failure is detected at this level, the corresponding entry in the neighbor set table is deleted if no other link is found in the meantime.

3.3.2.1.3. Declaration of selector MPRs

This process consists of informing the local node that it has been chosen as an MPR. In the HELLO message, neighbor type is initialized as MPR_NEIGH. On receiving the HELLO message, the node updates its MPR selector set, i.e. the group of nodes using it as an MPR.

3.3.2.1.4. HELLO message format

The format of HELLO messages is given in Figure 3.7. This message makes up the body of an OLSR message as shown in Figure 3.5. The packet is sent in the data field of the OLSR packet, with Message Type and TTL both set at 1.

0	1	2	3
0 1 2 3 4 5 6 7 8 9 0 1 2 3 4 5 6 7 8 9 0 1 2 3 4 5 6 7 8 9 0 1			
Reserved		Htime	Willingness
Link Code	Reserved	Link Message Size	
Neighbor Interface Address			
Neighbor Interface Address			
……..			
Link Code	Reserved	Link Message Size	
Neighbor Interface Address			
Neighbor Interface Address			

Figure 3.7. *HELLO message format*

The fields of the HELLO message are defined as follows:

– *Reserved*. This field must be set as 0000000000000 to conform to the specification.

– *HTime*. This field specifies the emission interval between HELLO messages used by the interface. If "*a*" is the four MSB and "*b*" is the four lsb, the emission interval is calculated as follows:

$$\text{HELLO emission interval} = C \times 1+\left(\frac{a}{16}\right) 2^b \text{ (in seconds)} \quad [3.2]$$

where the value of C depends on the application.

– *Willingness*. This field shows the willingness of a node to transmit traffic from other nodes. A node with "willingness=WILL_NEVER" should never be selected as an MPR. A node with "willingness=WILL_ALWAYS," on the other hand, should always be selected as an MPR. By default, a node should have "willingness=WILL_DEFAULT."

– *Link code*: This field contains information concerning the link between the emitting interface and the list that follows neighboring interfaces. It also contains information on the status of neighbors.

– *Neighbor type* uses values of {0, 1, 2, 3} to show, respectively, a lack of information on the link in question, a symmetric link, an asymmetric link, or a failed link.

– *Link type* shows if the link connects the node to a direct symmetric neighbor, an MPR, or not available.

7	6	5	4	3	2	1	0
0	0	0	0	Neighbor Type		Link Type	

Table 3.3. *Link code field format*

– *Link message size*: Given in bytes and measured from the beginning of the link code field to the beginning of the next link code field, or, if there are no more fields of this type, to the end of the message.

– *Neighbor interface address*: The interface address of a neighboring node.

3.3.2.1.5. MPR technique

The concept of MPRs aims to reduce the number of superfluous transmissions during the generalized diffusion of a message. MPRs are particularly useful in transmitting control messages over the network, optimizing the classic diffusion mechanism.

Each node in the network selects, independently of other nodes, its own group of MPRs based on knowledge of the neighborhood at two hops. The node should then inform those neighbors selected of their new role.

In a mobile environment with topology that changes unpredictably, as in the case of ad hoc networks, the set of MPRs has to be recalculated each time change is detected in the two-hop neighbor set. For this reason, the status of MPRs is set for a limited time by the neighborhood.

Figure 3.8 shows the improvement in diffusion methods engendered by the use of MPRs. In the first part of the illustration, a central node diffuses control messages to eight other nodes using the classic flooding technique.

On the other hand, if the relay technique is used, we notice that only four nodes are responsible for relaying messages. The choice of MPRs thus helps to reduce the number of redundant transmissions.

The choice of a group of MPRs appears simple. However, in reality, the choice of MPRs must be made very quickly, given the context of ad hoc networks, and with the smallest possible number of calculations.

We shall now present a heuristic for MPR selection. This method consists of receiving an MPR set which allows a node to transmit to all other nodes at two hops via MPRs. For this to be possible, the node must have both a neighbor set and a two-hop neighbor set.

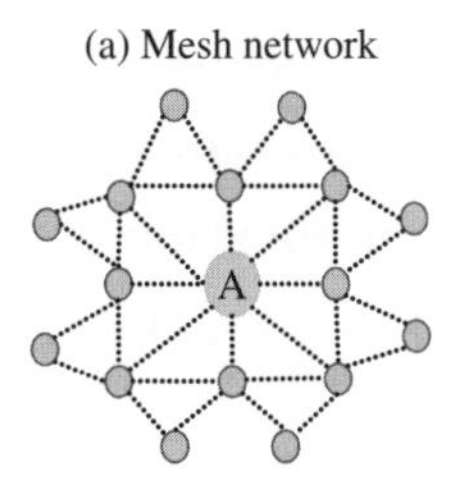

OLSR node

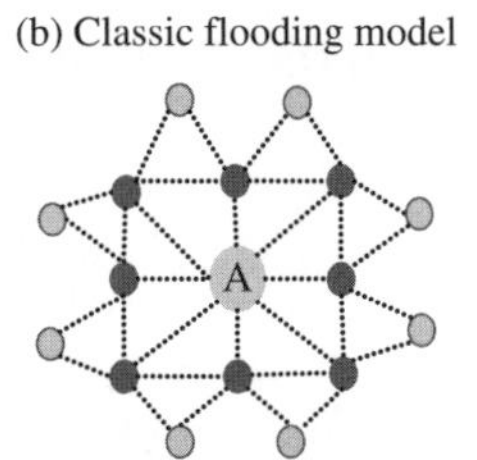

Direct neighbor affected byflooding

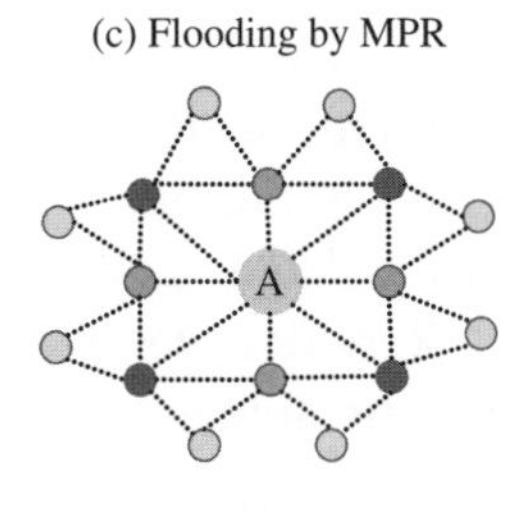

Node at 2 hops covered by MPRs
Nodes not involved in flooding
MPR node

Figure 3.8. *Flooding: classic model and via MPR*

We shall use the following terms:

– *N*. The group of direct neighbors of a node *x* through the interface *I*.

– *N2*. The group of level 2 neighbors accessible through the interface *I* excluding:

– nodes accessible via a member of *N(x)* with "Willingness=WILL_NEVER,"

– the node in question, x,

– all neighboring nodes with a symmetric link across one of the interfaces.

– *D(y)*. The degree of a node y of level 1 (y being an element of the *N(x)*) is defined as the number of symmetrically linked neighbors excluding all elements of *N(x)* and node x.

The algorithm for this heuristic is as follows:

Step 1. Begin with a group of relays containing all the elements of *N(x)* for which the ability to transmit traffic is equal to WILL_ALWAYS;

Step 2. Next, calculate *D(y)* for all y of *N(x)*;

Step 3. Add to *MPR(x)* the nodes of *N(x)* which are the only ones able to transmit traffic to an element of *N2(x)*. For example, if a node b, element of *N2(x)*, is only accessible over a symmetric link to a node a, then node a is added to *MPR(x)*. Level 2 nodes covered by the group *N2* are then eliminated from the selection;

Step 4. As long as there is still a node in *N2* not covered by at least one MPR, i.e., $N2(x) \neq \{\}$, proceed as follows:

a) for each node of *N*, calculate the reachability, i.e. the number of nodes in *N2(x)* not covered but accessible by this node,

b) select the node with the highest N_WILLINGNESS value from the group of nodes *N* with an MPR value other than zero, add it to MPR(x) and remove all nodes covered by this node from *N2(x)*;

Step 5. The MPR set of a node is found by uniting the MPRs of different interfaces.

An example of an MPR selection mechanism is given in the following figure.

Node	Neighbors at one hop	Neighbors at two hops	MPR
B	A, C, F, G	D, E	C

Table 3.4. *MPR selection in the OLSR protocol*

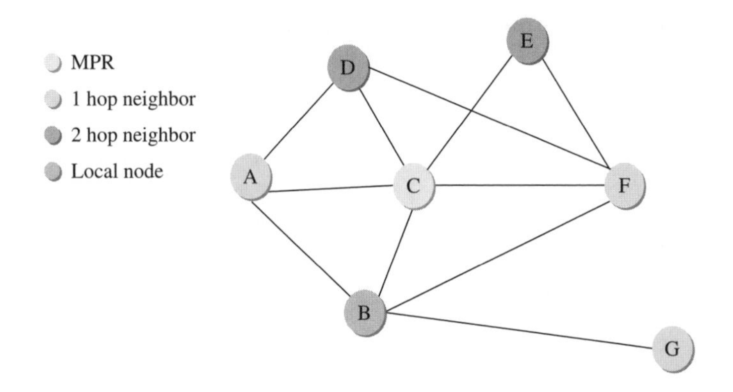

Figure 3.9. *Example of MPR selection*

Node *B* selects *C* as relay point as this enables *B* to reach its neighbors at two hops (*D* and *E*), *{D(C) = 5, D(A) = 2, and D(F) = 3}*. The objective of a relay point is always to reach more distant neighbors.

3.3.2.2. *Topology management*

Topology management in the OLSR protocol is carried out using control messages, which allow nodes both to detect neighbors and select MPRs that will allow them to diffuse control information in the network.

If a node possesses several interfaces, control information must be sent to all these interfaces. Paths are then constructed using MPRs and direct links to neighbors.

3.3.2.2.1. TC messages and the topological information base

TC messages are sent periodically by each MPR. They contain a list of neighbors that are using the node in question as an MPR. This information is important, as paths are constructed from MPRs, so if a node is to be reached, it is essential to know which nodes constitute its MPRs.

The analysis of TC messages allows each node to maintain a topology base for every destination. TC messages are diffused across the whole network using OLSR's optimized diffusion method via MPRs.

0	1	2	3
0 1 2 3 4 5 6 7	8 9 0 1 2 3 4 5	6 7 8 9 0 1 2 3	4 5 6 7 8 9 0 1
ANSN		Reserved	
Advertised Neighbor Main Address			
Advertised Neighbor Main Address			
Advertised Neighbor Main Address			
...			

Figure 3.10. *TC message format*

The fields of TC messages are defined as follows, according to RFC 3626:

– ANSN. *Advertised Neighbor Sequence Number* – a sequence number associated with the group of neighbors receiving the message, indicating the freshness of the information.

– *Reserved.* This field is reserved and must be fixed at 0000000000000000.

– *Advertised neighbor main address.* This field contains the address of the neighboring node averted by topological changes.

3.3.2.2.2. MID messages and the declaration of multiple interfaces

Each node detects every interface of its neighbors, which possesses a direct symmetric link to one of its interfaces. The propagation of radio waves is subject to perturbations that can render links between two nodes asymmetric. To avoid these scenarios, links are checked in both directions before being considered as valid.

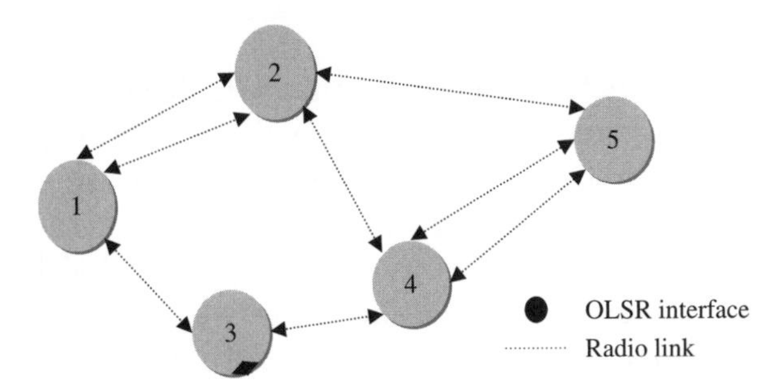

Figure 3.11. *Presence of multiple interfaces in OLSR nodes*

A node with several interfaces periodically sends an MID message containing its address and the addresses of its interfaces (the diffusion of which is optimized using MPR).

3.3.2.2.3. Management of subnetworks

A participating node in a MANET can also be connected to other types of networks, for example a wired network. In this case, the node can operate as a gateway offering access to different sources from one type of network to another. In order to do this, the node must be able to inform the rest of the MANET that it is possible to contact another network, or another host, via this node.

This scenario is different to that concerning the declaration of interfaces associated with a node. All the interfaces declared in MID messages are part of the MANET, where the nodes use a routing protocol such as OLSR. In the case currently being discussed, interfaces participate in another non-MANET.

To give access to these networks and/or hosts from the other side of the MANET and to transmit the information that another kind of network or host may be contacted through the node in question, each node in the MANET must maintain a group of *tuples* for each node or gateway.

3.3.2.3. *Routing*

It is important to understand that OLSR is not a "routing protocol" *per se*, as it is not responsible for the process of routing. It is better described as a "route maintenance protocol", as it maintains routing tables for the transmission of data. Under no circumstances does the protocol create a list of nodes through which a packet must pass to arrive at a destination. Instead, it updates the routing table of the core of the operating system. Thus, in spite of these precisions, OLSR is still referred to as a routing protocol.

The routing of data, in the figurative sense of the term – a better phrase would be the transmission of data – is done hop by hop on the basis of information collected from control packets received.

Control packets themselves are diffused across the network via MPRs. A node retransmits a message if the message is received from a node for which it is the MPR and has a lifespan (TTL)[1] greater than one, after checking that the message in question has not already been received.

Each node maintains a routing table, which allows it to route data to any destination present in the network. The

---

1 The lifespan, or Time to Live, of a packet is the maximum number of hops a packet can travel. TTL is reduced by intermeditory nodes each time the pocket is retransmitted.

calculation of the routing table is based on information contained in the neighborhood information base and the topological base in combination with interface associations. For this reason, each time one of these addresses changes, the routing table is updated and recalculated. The routing table follows the format shown in Table 3.5.

R_dest_addr	R_next_addr	R_dist	R_iface_id
R_dest_addr	R_next_addr	R_dist	R_iface_id

Table 3.5. *OLSR routing table format*

R_next_addr is the identifying address of the node at a distance of one hop, which must be passed through to get to the node identified as R_dest_addr. R_iface_addr is the identifier of the local interface through which the node may reach R_dest_addr. R_dist is the estimated distance, in hops, separating R_dest_addr from the local node. The entries in the routing table correspond to all nodes for which the local node can calculate a valid route. Other destinations, for which only a partial route is known, or where a broken link is involved, are not saved in the routing table.

The routing table is updated every time a topological change is detected in the topological information base, more precisely when a *tuple*[2] appears or disappears from the topological base. This update does not result in the generation of messages in the network, merely a local recalculation.

The calculation of a routing table is carried out using the following procedure:

– Delete all previous table entries.

---

2 A tuple is a table entry, whether in a table of neighbors, a routing table, etc.

– Insert all direct symmetric neighbors within 1 hop into the table. This information is extracted from the neighbor base. For each entry, add: R_dest_addr, which corresponds to the principal address of the neighbor L_neighbor_iface_addr in the neighbor base; R_next_addr, which corresponds to the neighboring interface L_neighbor_iface_addr; and R_iface_addr, which corresponds to L_local_iface_addr. R_dist naturally begins at 1.

– Insert destinations further away than one hop ($h + 1$). The following procedure is conducted for every value of h, starting at $h = 1$ and increasing h by 1 each time. This loop stops when there are no more new entries in the routing table.

– For every *tuple* in the topological information base, if T_dest_addr does not correspond to an R_dest_addr in the entries in the routing table, and T_dest_addr corresponds to one of the R_dest_addr with R_dist equal to h, then a new entry is made in the routing table, such as:

- R_dest_addr = T_dest_addr,

- R_next_addr = R_next_addr of the entry for which R_dest_addr is equal to T_last, and

- R_dist = h+1.

– The routing table is completed using the group of interface associations, adding entries for all interfaces not present in the routing table. Finally, routes are inserted towards the group of network associations and attached hosts. Evidently, the address of the next hop and the distance are the same as those of the principal address allowing access to these interfaces and subnetworks.

3.4. Simulation environment

Several network simulation programs are currently in existence, of which OPNET and Network Simulator are the best known.

The first of the two, with a very user-friendly interface, is the simplest to use; unfortunately, OPNET is not in open access. The only available version – the IT Guru version of OPNET – is very limited from a simulation point of view.

The IT Guru version of OPNET does not allow for the simulation of ad hoc networks. Only the "Modeler" version of OPNET, commercialized under license, includes AODV, DST, OLSR, etc.

For this reason, we have chosen to run our simulation under ns simulator.

3.4.1. *The ns-2 network simulator*

The ns simulator has become a reference standard. The ns-2 simulator is a research tool used in the design and comprehension of networks. It permits the user to develop a network and simulate communications between nodes.

The ns-2 simulator was developed by the VINT research group at the University of Berkeley. The Monarch research group at Carnegie Mellon University (http://www.cmu.edu) extended the simulator to deal with node mobility, with a realistic physical layer including propagation models and a MAC layer that implements the IEEE 802.11 interface, using the distributed coordination function (DCF) as a means of access. The network interface card (NIC) network interface card uses values specified by Lucent's WaveLan cards. The model takes account of collisions, propagation delays, and the extenuation of the signal. Each node can transmit over a maximum of 250 m.

The simulator uses the object-oriented language OTCL, derived from TCL, to describe simulation conditions in the form of scripts. The user supplies the network topology and the characteristics of the physical links, the protocols being used, the type of traffic generated by the sources, etc.

Although the OTCL script makes the simulator easy to use (e.g. for editing or modifying simulations), the basic script is written in C++ for increased calculating power. A large number of classes are predefined and implement several types of protocols, waiting lists, sources, and routing algorithms.

Simulations produce text files containing all the events of the simulation. Treatment of this file then allows the required information to be extracted. In addition, the simulator allows an animation file to be created enabling visualization of the simulation using the network animator (NAM) graphics interface. The visualizer provides a visual representation of the network map, through which we can observe packets circulating, monitor queues, etc.

The ns-2 simulator implements certain ad hoc protocols including DSR, DSDV, and AODV. However, the OLSR protocol is not included. There are only two available versions of this protocol. One, published by the INRIA,[3] known as OOLSR, is an object-oriented implementation of the protocol in version ns-2.27. A second version of OLSR is available for version ns-2.28.

3.4.2. *Methodology*

We shall simulate a network of 500 × 500 m with a variable number of nodes (10 or 30). We aim to observe the effect of network densification on the performance of each of the two protocols under consideration, as the two react differently to this effect and to the speed of nodes.

---

3 Institut National de Recherche en Informatique et en Automatique.

For the physical layer, we have simulated an omnidirectional antenna (OmniAntenna) and a card, which conforms to IEEE 802.11 norms. Thus, each node has a range of 250 m in the absence of obstacles, and a nominal bandwidth of 2 Mbit/s.

The radio propagation model being used is the two-ray reflection model. This model takes into account the reflection of radio waves on the earth, and has been shown to provide sufficiently accurate predictions over long distances. The signal degrades at a rate of $1/d^4$ where d is the distance from sender to receiver.

Each node has a queue with a maximum capacity of 50 packets.

The traffic model is as follows: for every scenario, we have simulated constant bit rate sources. These traffic sources modelize the application layer and are located on UDP transport agents, which modelize the transport layer. The maximum number of connections to be established is fixed at 6. Each source sends out packets with a size of 512 bytes and a flow rate of five packets per second. By setting traffic at these levels, we hope to observe performance in a network where individual nodes have constant traffic. The rate is 4 Kbit/s.

The mobility model used is called the *Random Waypoint Model*. In our simulations, nodes are initially placed at random over a surface of 500 × 500 m. Each node then chooses a random direction from 0 to 360° and a speed lower than Vmax (between 0 and 20 m/s inclusive), at which it moves in the chosen direction.

Upon arriving at the edge of the network, it becomes immobile for a time (pause time). Once this time has elapsed, the node chooses a new direction, this time between 0 and 180°, depending on the border the node has reached. This process is repeated until the end of the simulation. For our simulations, the pause time was set at zero.

Two parameters can therefore affect node mobility within the network: Vmax speed and the maximum pause time. We have chosen to vary the speed of movement while maintaining the pause time as zero.

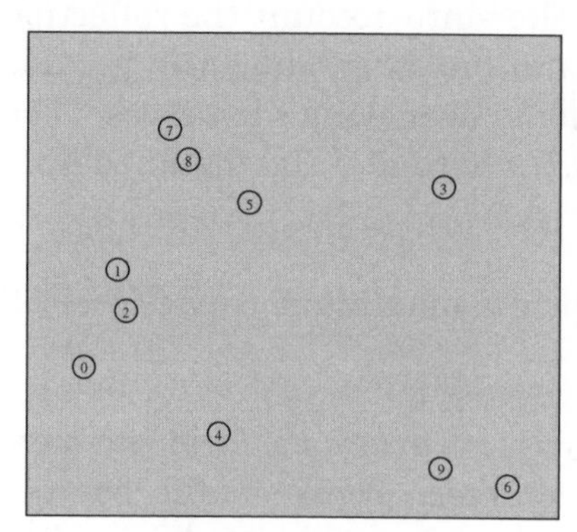

(a) 500 m × 500 m, ten nodes

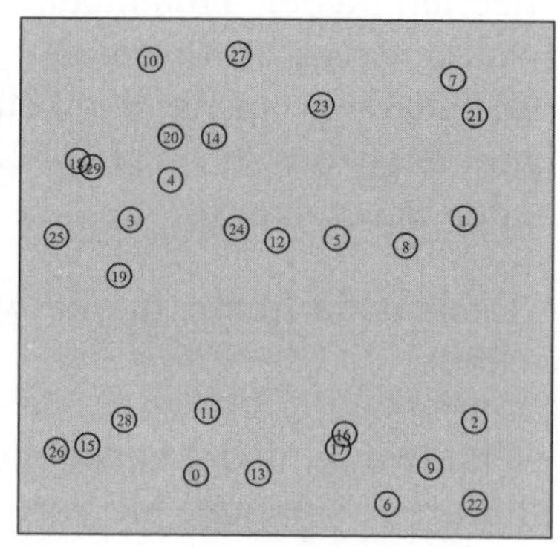

(b) 500 m × 500 m, thirty nodes

Figure 3.12. *Topology of the simulated network*

3.4.3. *Parameters to evaluate*

3.4.3.1. *Average packet delay*

The main aim of all protocols is to offer guarantees to applications concerning the time taken to transfer data packets from end-to-end; thus, we first evaluate this parameter for the two protocols (AODV and OLSR) and compare the results.

This parameter gives the average time necessary to transfer a data packet from one end of the other. It is calculated as follows:

$$\text{Average packet delay: APD-1000}\frac{\sum[T_{AR}(i)-T_{AS}(i)]}{\sum \text{packets received}}\text{(in ms)} \qquad [3.3]$$

where $T_{AR}(i)$ is the instant when the data packet i is received by the destination transport agent and $T_{AS}(i)$ is the instant when the data packet i is sent by the source transport agent.

3.4.3.2. *Packet delivery success rate*

When evaluating a protocol, it is very important to consider the capacity of the protocol to transmit data packets to their destination.

This parameter is the percentage of packets sent within the network which are successfully delivered to destination:

$$\text{Packet delivery ratio: PDR} = 100\frac{\sum \text{packets received}}{\text{packets sent}} \text{ (in \%)} \quad [3.4]$$

3.4.3.3. *Traffic overhead (TOH)*

As bandwidth in an ad hoc network is a limited and shared resource, maximum economy is necessary.

Based on this principle, every proposed routing protocol must be evaluated in terms of the volume of traffic generated, which is important in determining the protocol's consumption of bandwidth.

For the AODV protocol, the traffic overhead is calculated as follows:

$$\text{Traffic overhead} = \text{TOH} = \sum \text{RREQ, RREP} \text{ (in pkts)} \quad [3.5]$$

For the OLSR protocol, this metric is expressed by equation [3.6]. In this case, HNA packets are not counted as we consider that the network is not connected to an external non-MANET.

$$\text{Traffic overhead} = \text{TOH} = \sum \text{HELLO, TC, MID} \text{ (in pkts)} \quad [3.6]$$

3.4.3.4. *Route acquisition latency (RAL)*

This parameter provides information on the average time required by the routing protocol to find a route to a destination, as we are dealing with a case of on-demand routing. It also gives an indication of the time a data packet spends in the queue of a node before being sent over the network.

$$\text{Route acquisition latency: RAL} = 1000\frac{\sum_i [Ts(i) - Tr(i)]}{\text{packets sent}} \quad \text{(in ms)} \tag{3.7}$$

where $Ts(i)$ is the moment when the data packet i leaves the router and $Tr(i)$ is the instant when the data packet i is delivered by the transport agent to the network layer.

3.5. Results and analysis

3.5.1. *Packet delivery ratio*

It should be noted that the packet delivery ratio (PDR) is the measurement used to demonstrate the reliability of a protocol; after all, the principal function of a routing protocol is the transmission of information.

At first view, Figure. 3.13 shows a reduction of PDR as speed increases for both protocols in both scenarios. This behavioral aspect as speed increases is due to broken links. An increase in mobility in a network causes on the one hand a rise in the number of broken links and on the other an increase in the length of routes from source to destination. This has the effect of increasing propagation time for supervision messages between the moment the link breaks and the moment the source node is informed of the fact. All packets sent across this broken link in the meantime will be lost.

The AODV protocol is less sensitive to this effect than the OLSR protocol, and the PDR is above 95%. This rate is more than sufficient for communications without retransmission mechanisms in the case of packet loss. OLSR performs less in high-mobility conditions. In a static network, AODV and OLSR have practically the same PDR (~99%). The OLSR protocol maintains up-to-date routing tables. The selection of MPRs is therefore efficient. TC and HELLO control messages reach their target and allow for an improved knowledge of network topology and link-state.

However, once mobiles start to move, these messages may not reach their destination. The rate of packet loss increases by approximately 4% with each 5 m/s increase in speed. At a speed of 20 m/s, packet loss can reach 12%; at the same speed, AODV has a success rate of 98%.

This can be explained in three ways:

– The absence of collision detection mechanisms when using the IEEE 802.11 standard, which provokes high levels of collisions between packets, with a negative effect on knowledge of network topology. This effect is less noticeable in AODV, as packet transmission does not depend on stored information (i.e. tables) but on path finding as and when required. The AODV protocol is therefore better able to cope with collisions and broken links than OLSR.

– Another explanation of the phenomenon whereby PDR diminishes rapidly as speed increases is the nature of radio links and the limited range. If a mobile unit moves at 15 m/s, it will be outside the mobile transmission zone in around 14 s.

– Another cause of data loss with increased speed is the limited size of queues allowed by nodes (50 packets) in relation to the amount of traffic sent within the network (100,000 packets per source). Several packets of varying types may want to use or reach the same node. The number of such packets can be higher than the capacity of the queue, i.e. any further packets will be rejected. This phenomenon is accentuated in OLSR, as the MPRs are solely responsible for transmitting TC messages. The rejection of TC messages by these nodes leads to an imperfect knowledge of network topology, thus higher rates of packet loss. Similar behavior is observed in relation to network densification.

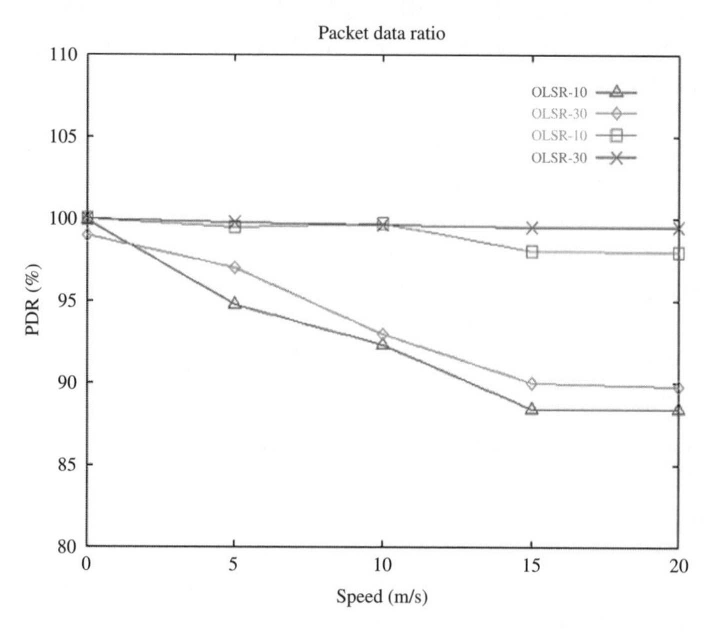

Figure 3.13. *Comparison of PDR for the AODV and OLSR protocols*

As density varies from 10 to 30 N/km^2, the level of packet loss is reduced for both protocols. Effectively, the more nodes there are in a network, the more routing possibilities exist. In a network with fewer nodes, it is more probable that a node will be out of the range of all the others, rendering it inaccessible. This effect is more noticeable for the OLSR protocol than for the AODV. The PDR for a network of 30 nodes is constantly higher than that for a ten-node network. The denser the network, the more effective is the OLSR protocol.

3.5.2. Average packet delay

Figure 3.14 allows us to consider the average time taken for any given packet to travel from source to destination. This end-to-end time only applies to packets successfully delivered to destination. Packets lost *en route* are not taken into consideration in the calculations.

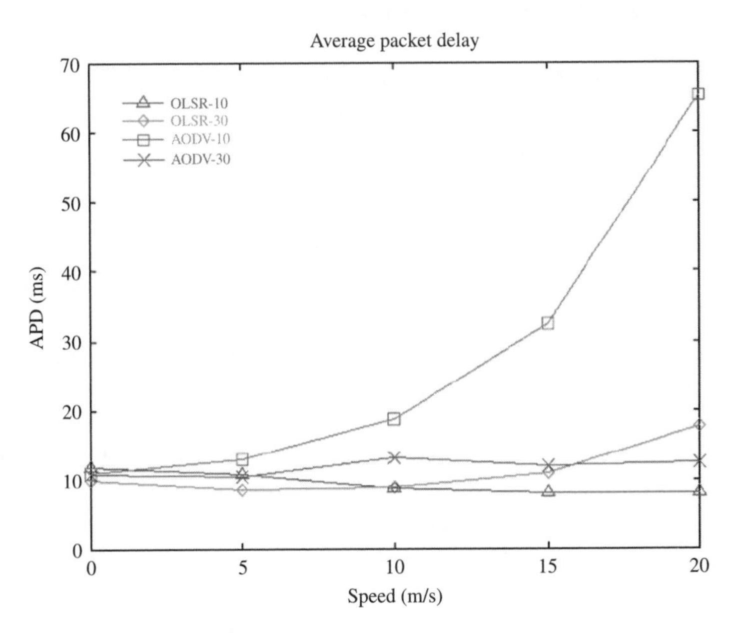

Figure 3.14. *Comparison of average packet delay for AODV and OLSR protocols*

In the case of a static network ($V = 0$ m/s), the time taken to transmit a routed packet using the OLSR or the AODV protocol in a best-routing version is around 10 ms. However, the time delay is considerably greater in a ten-node network using the AODV protocol.

Effectively, the more spread out the nodes are, the lower the possibility of finding a route. If nodes then begin to move, routes become invalid and a new route-seeking procedure must be launched. The time delay involved increases by 10% with every 5 m/s increase in speed, reaching 65 ms for a speed of 20 m/s. However, if the network is denser, the probability of successful data transmission is higher as, even with high mobility, routes are easier to find.

3.5.3. *Control traffic volume*

The control traffic generated by a routing protocol is an interesting parameter to measure as it shows clearly the difference between proactive and reactive protocols as well as determining bandwidth consumption.

It is important to note the huge difference between the control traffic generated by the two protocols in question. OLSR generates on average 13,000 packets for a network of 30 nodes. AODV, on the other hand, generates around 1,800 packets for 30 nodes.

For the OLSR protocol, traffic comprises HELLO, TC, and MID messages. The size of these messages increases with network density. AODV uses HELLO messages, sent periodically, and RREQ and RREP messages for pathfinding.

We notice an almost linear variation in control traffic in relation to speed. The traffic overhead reaches 12,800 packets for a network of 30 nodes, but only one-third of this value for a network with one-third of the density. This is to be expected as the number of control messages sent is a function of the number of nodes in the network.

In fact, the OLSR protocol does not possess a retransmission mechanism for control messages as and when these messages are not delivered. The generation of supervision messages is regulated solely by a periodic message. Time intervals specific to each control message are fixed for the sending of these periodic control messages.

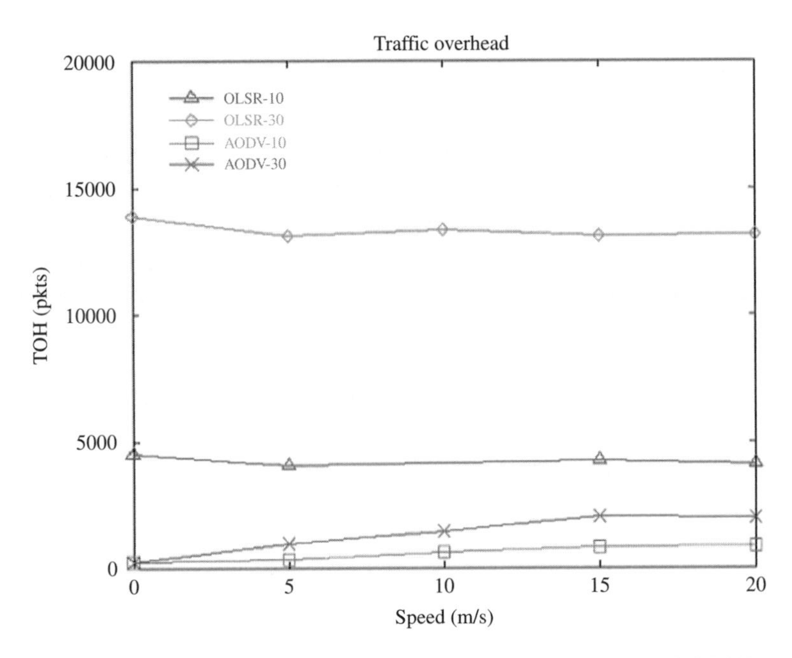

Figure 3.15. *Comparison of traffic overhead for AODV and OLSR*

For the AODV protocol, however, the TOH variation is not linear; it grows with the speed and the density of the network. This can be explained by the fact that the AODV protocol uses additional control packets to deal with topological change and find appropriate routes.

3.5.4. *Route acquisition latency*

This parameter is particularly important for applications where data must be delivered as quickly as possible. For the OLSR protocol, route establishment time is not a consideration, as the protocol maintains various tables containing topological information that allow routes to be found at any moment, without recourse to a pathfinding mechanism. However, AODV requires a route discovery process to be started each time a new route is needed, which takes a certain time.

The AODV protocol therefore takes a longer time to establish a valid route. This effect is more noticeable in low-density networks. The protocol sends out several requests before finding a valid route. This RAL increases noticeably as speed increases for the same network. The AODV protocol has difficulties in maintaining valid routes over long time periods in circumstances of high mobility and low-network density. The higher the number of nodes in the network, the more valid routes there will be.

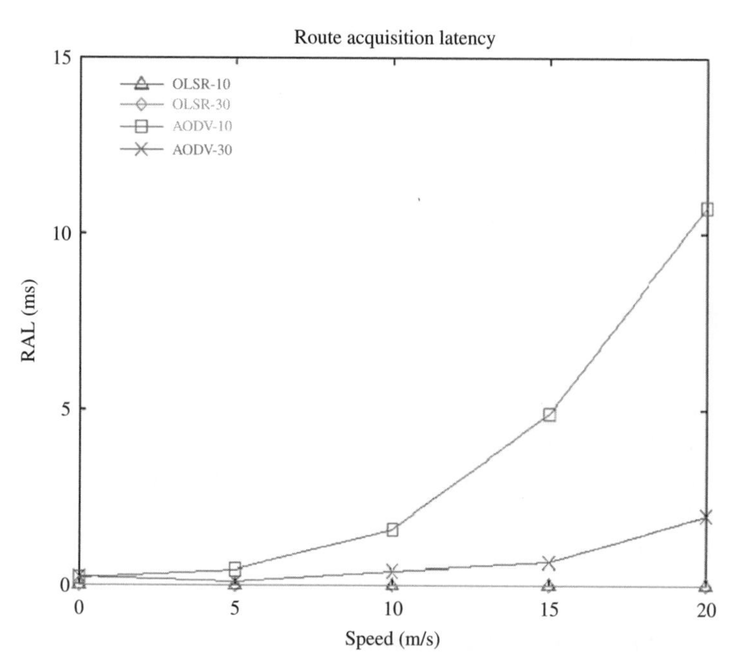

Figure 3.16. *Comparison of route acquisition latency in AODV and OLSR*

3.6. Conclusion

In this chapter, we have attempted to compare two protocols from different families, OLSR and AODV, varying different parameters (speed and network density). The results obtained suggest that AODV should be favored over OLSR if the primary aim is to guarantee the delivery of packets. Nevertheless, the proactive (table-driven) protocol is faster if we consider the end-to-end time delay for packet delivery. The maintenance of route information, as practiced by OLSR, constitutes an advantage in guaranteeing a rapid response to a route request. We have demonstrated that, generally, the denser the network, the easier it is to transmit traffic. Nevertheless, the traffic overhead is a constraint for the OLSR protocol.

In the following chapter, we shall present some solutions on the state of the art of QoS in ad hoc networks in order to later present a solution for supporting QoS in the OLSR protocol.

Chapter 4

Quality of Service in MANETs

4.1. Introduction

Given the dynamic topology of mobile ad hoc networks (MANETs) and the limited bandwidth available, supplying quality of service (QoS) is a major challenge. Several research projects are currently underway with the aim of providing QoS on the Internet and in other network architectures, but most of the proposed solutions are incompatible with ad hoc networks. The study of QoS for ad hoc networks is important, as QoS would enable more sensitive network administration and more equal resource sharing, optimizing network costs.

For QoS to be supported, link information such as delay, available bandwidth, cost, and error levels needs to be available and controllable. However, it is difficult to obtain this kind of information in ad hoc networks as the radio link changes according to environmental circumstances. Limited resources and node mobility complicate the task still further.

The standardization of routing protocols for ad hoc networks has aroused growing interest among research groups in the field, which aim to improve performance and satisfy application constraints. A number of drafts have appeared on the subject, notably from the MANET, HIPERCOM, and Nokia searching center (NSC) groups. They deal with different ad hoc protocols in great detail, including ad hoc on-demand distance-vector (AODV), dynamic source routing, optimized link-state routing (OLSR), and TBRPF. Multimedia applications usually require QoS guarantees and the reservation of resources. This has led to the development of a number of techniques to improve routing. The OLSR protocol, a link-state routing protocol developed by the Institut National de Recherche en Informatique et en Automatique, does not include any specification concerning data transmission guarantees.

From this development has come the idea of developing extensions for the OLSR routing protocol, which would give the possibility of control and respond to application requirements. Thus, through routing with QoS, the proactive OLSRQSUP protocol goes beyond the demands of best effort routing to provide QoS guarantees (constraint routing). To this effect, we have developed a protocol to be known as OLSRQSUP, based on OLSR but with new metrics that deal with delay and bandwidth.

In this chapter, we first give an overview of the state of the art of QoS notions in ad hoc networks. We shall then explain the operating mechanisms of our new protocol, implemented in the ns-2 simulator, with the extensions and modifications made to the basic OLSR protocol.

4.2. QoS: a definition

"QoS" is a term CCITT[1], which has been much used of late in the domain of wired networks. The exact definition of the term has been a cause of debate. Recommendation E.800 of

1 Consultative Committee for International Telephony and Telegraphy

the defines QoS as follows: “the collective effect of service performance which determines the level of satisfaction of the service user”. This definition is insufficient, as it makes no reference to important characteristics such as delay and bandwidth, and so requires modification.

QoS is defined as “the group of guarantee parameters which define the behavior of a network under certain conditions”. These parameters are the subject of an agreement between the client and the network operator or service provider. This definition works for wired networks such as ATM and X25, where a centralized architecture means that only the network administrator is responsible for guaranteeing QoS and for monitoring any infringement of the terms of the contract. However, this is not the case in ad hoc networks, which have no fixed infrastructure or centralized administration. For this reason, we refer to “QoS-oriented routing”, rather than just “QoS”, in discussing ad hoc networks.

QoS in routing is defined as “the requirements which must be satisfied by the network during the transport of flux from a source to a destination”. The network must therefore guarantee a certain number of measurable parameters that characterize a well-defined service to satisfy constraints from end-to-end, such as delay, jitter, flow rate, and packet delivery ratio (PDR). This allows treatments to be adapted to applications, more precise network administration, more equitable distribution of resources, and an optimization of network costs.

4.2.1. *QoS in wired networks*

Best effort routing, as offered by Internet protocol (IP) networks, is sufficient for so-called “elastic” applications such as e-mail and file transfers. However, multimedia applications need stronger guarantees. For this reason, the

Internet engineering task force (IETF) has successively developed two architectures: integrated services (IntServ) and differentiated services (DiffServ).

4.2.1.1. *The IntServ/RSVP approach*

The IntServ/resource reservation setup protocol (RSVP) model is a solution proposed by the IntServ workgroup at the IETF, allowing Internet-style networks to be transformed into networks with IntServs, i.e. networks capable of dealing with multimedia flux and transporting classic traffic. Using IntServ/RSVP, applications can reserve resources along a point-to-point or multipoint network. Reservations are made for each data stream, i.e. for each application. With this aim, each router needs to be able to classify packets to determine which stream they belong to, authorize the transmission of packets according to reserved and available resources and monitor the state of resources before accepting or declining a reservation request. The IntServ model uses the RSVP signaling protocol to establish reservations.

The principle criticism of the IntServ model is that it is not particularly suitable for scaling up. The management of resources by stream effectively requires several states to be saved and maintained by each router. In a small network, such as a local network, the cost of this kind of mechanism is relatively limited, but in high-density networks, such as the Internet, the IntServ model is too costly to be applied. The extra cost in terms of treatment required from each router in an interconnecting network is too high and would have a noticeable negative effect on the performance of the network as a whole.

4.2.1.2. *The DiffServ approach*

The DiffServs approach is based on the aggregation of data streams into flux classes depending on their requirements in terms of QoS. Each data stream specifies its QoS needs using a field in the IP header. Depending on the value present in this field, a particular treatment, PHB, (per hop behavior) is applied by each router through which the data travels, enabling the required QoS to be supplied.

The network is divided into DiffServ domains interconnected by border routers to form a DiffServ region. A terminal node (sender or receiver) is connected to a DiffServ domain by the intermediary of a border router. These border routers have a special role to play. On receiving data from a sender, they must classify and package data to make it conform to the requested QoS. At the border of two DiffServ domains, they must modify incoming traffic to adapt it to the service available in the domain it is about to cross.

4.2.2. *QoS in wireless networks*

The capacity of an ad hoc network to supply QoS depends on all component parts of the network, from the physical layer to the multiple access collision (MAC) layer and the network layer. Figure 4.1 provides a visual representation of QoS protocols in ad hoc networks. Various approaches to this problem and related studies are presented in section 4.2.2.1.

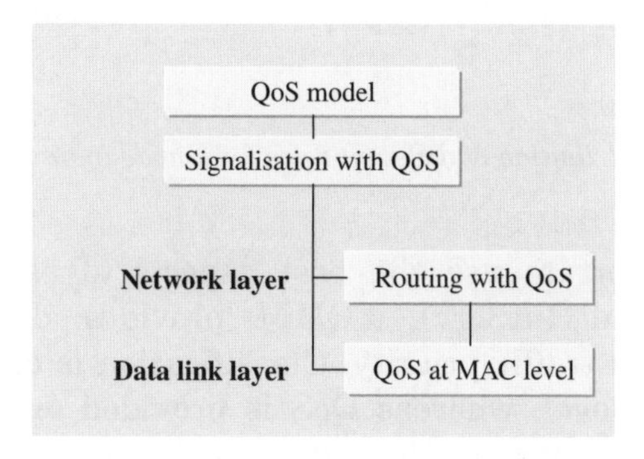

Figure 4.1. *QoS in ad hoc networks*

4.2.2.1. *QoS models*

Flexible QoS Model for MANET (FQMM) is based on a QoS model conceived specially for MANETs. This model combines the mechanisms of IntServ and DiffServ. The FQMM offers a hybrid schema adding the "by stream" management of IntServ to the "by class" option provided by DiffServ, with a relative and adaptable traffic profile used to maintain differentiation between types of traffic and network dynamics. Each node uses a controller that manages the profile of each item of traffic (as a function of the available bandwidth), marks packets, and regulates traffic.

FQMM uses any routing protocol capable of providing routes corresponding to a QoS demand. When a node wishes to send, nodes on the path from source to destination are considered as intermediary nodes that will transmit packets.

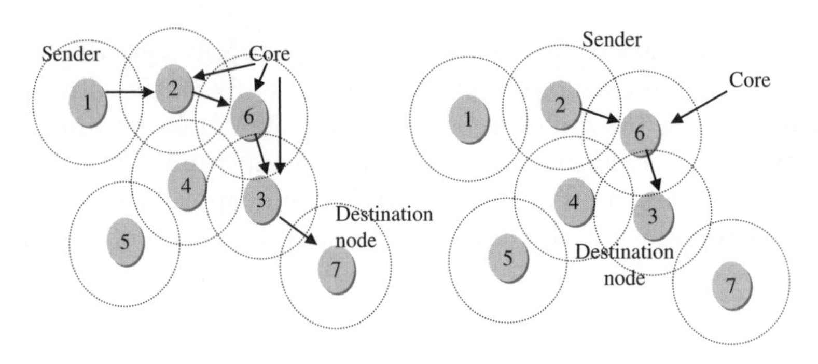

Figure 4.2. *Attribution of different roles for a node in an ad hoc network*

In an effort to use the best aspects of each approach (DiffServ and IntServ), FQMM provides different QoS depending on traffic priority. Classification is carried out at source-node level, whereas QoS is provided over the whole length of the journey. However, problems linked to ad hoc network operation are left to be dealt with by the underlying protocols.

4.2.2.2. *Signaling*

Above the link layer, QoS signaling operates as a QoS control center supporting QoS. The function of the reservation QoS is defining a connection from source to destination, involving the reservation of resources in intermediary nodes. The reservation QoS reserves resources, installs, destroys, and renegotiates flux in the network. Signaling can be done inside data packets (*in-band* signaling) or through separate control packets (*out-band* signaling).

At the time of writing, no out-band protocol has been proposed. Only the INSIGNIA in-band signalization protocol permits bandwidth reservations to be carried out in MANETs. INSIGNIA supports QoS in MANETs.

When an application wishes to make a bandwidth reservation, it sends a first data packet to the receiver in *best effort* mode, with a reservation request in the header. The application continues to send packets of this kind until it receives notification that the reservation has been accepted. Every reservation request contains two values: the minimum bandwidth, below which the flux does not wish to descend, and the ideal bandwidth that would guarantee optimal functioning of the application. Each router on the path may choose to accept the reservation at maximum bandwidth, at minimum bandwidth, or with degradation of the flux to the best effort mode. The most restrictive routing policy chosen on the path will be used. Once the route is established, each data packet will contain a field in the IP header indicating the state of the flux.

Thus, during the transfer, each router along the path may decide to degrade the reservation to minimum bandwidth or increase it to maximum value by modifying the header of a data packet. The receiver sends regular reports to the sender on the state of the reservation, allowing the sender to modify the outgoing flow. This protocol causes minimal additional costs for signalization and, moreover, guarantees rapid route reconstruction. In this way, INSIGNIA supports rapid flux reservations, restoration, and storage algorithms developed to transmit real-time traffic in ad hoc networks.

4.2.2.3. *Routing with QoS*

4.2.2.3.1. Definition of routing with QoS

Before entering into a proper discussion of the subject, we should provide a proper definition of the term "routing with QoS".

Best effort routing usually consists of finding the shortest path (in terms of distance or delay) from a source to a destination for the purposes of data transfer.

In routing with QoS, the aim is not simply to find the best path according to certain criteria, but to find the best possible path overall.

To do this, a certain number of constraints are applied to routes to determine their eligibility. For example, one may wish to find a route with a certain level of available bandwidth to transmit video traffic, or a route that will guarantee packets will be received by the destination within a certain time of being sent by the source. Every route responding to certain quantitative criteria can be defined as a route guaranteeing a certain QoS.

Finally, it is also important to ensure optimization in the routing process. This can take the form of maximizing usage of the available bandwidth in the network or by serving the highest possible number of clients. Once again, it is important to point out that although routing with the lowest time delay (based on link-state) gives good results in wired networks operating on a best effort routing principle, this is not obviously the best choice for an ad hoc network.

Routing with QoS is therefore the process of establishing and maintaining optimal routes (for the communicating pair of nodes as for the network) satisfying certain criteria concerning the quality of data transmission. Although this objective is easy enough to reach in wired local networks, ad hoc networks have a number of specificities, which render the conception of this kind of algorithm difficult.

A routing protocol with QoS must operate using optimal signaling to find a route, which satisfies end-to-end constraints in QoS terms. A major difficulty for routing protocols with QoS lies in the fact that the traditional meaning of "an available route" is no longer valid in a QoS context.

Various approaches have been developed with the aim of introducing QoS into ad hoc networks.

4.2.2.3.2. Ticket-based probing (TBP) algorithm

One example of a protocol with QoS is TBP. The main idea is to use a ticket system to limit the number of nodes seeking routes at the same time. When a source wants to find a path with QoS, it simply needs to send messages with tickets.

Route searches are not carried out by flooding, but using the following method: a node, receiving a route request (RREQ) message, is permitted to transmit this message to a limited number of neighbors determined by the value of the ticket. The ticket number is determined according to information on the state of the node. A message carries at least one ticket and the maximum number of tickets limits the maximum number of routes. When an intermediary node receives a probe message with n tickets, and depending on the information it holds, it decides where to send the message. A protocol of this kind allows the scope of searches for routes satisfying a certain QoS to be limited.

4.2.2.3.3. CEDAR

In routing protocols with QoS, the aim is to find a route that obeys QoS criteria, often based on bandwidth or delay.

The main problem involves the additional cost generated, as the maintenance of route state information within the network and exchanges of information consume additional bandwidth in a system where bandwidth is a limited resource.

Another problem is the difficulty of obtaining precise and coherent route state information, made difficult by the dynamic nature of ad hoc networks.

CEDAR is a distributed routing algorithm with integrated QoS. This protocol does not aim to offer strict guarantees concerning bandwidth, but rather to allow each flux to obtain a bandwidth averaging the same as the bandwidth requested. It uses the notion of a dominant group (or network core).

Used in small networks, CEDAR is built of three main components:

– *Network core extraction*. A group of nodes is chosen dynamically to calculate routes and maintain the state of links in the network. The advantage of this approach is that, with a reduced group of nodes, information exchanges concerning link-state and routes are minimized. This reduces the number of messages circulating within the network. In addition, when a route changes, only the core nodes are involved in calculating the new route.

– *Link-state* propagation. Routing with QoS is carried out thanks to the propagation of information on stable links with high bandwidth.

– *Route calculation*. It is based on the discovery and establishment of the shortest path to a destination with the required bandwidth available. “Emergency” or “spare” routes are used if and when the principal route requires reconstruction. Reconstruction can be local (at the point where the break occurs) or at the initiative of the source.

Instead of calculating a route with the lowest possible number of hops, CEDAR aims to find a stable path to guarantee maximum bandwidth.

4.2.2.3.4. QoS AODV

Another approach to routing with QoS has been developed, in draft form, by the IETF with the aim of adding QoS parameters to the APDV protocol.

The work carried out consisted of adding QoS extensions to RREQ and route reply (RREP) messages, specifying required bandwidth and time delay in the route-seeking phase. A node can become a hop on a route only if these demands, as specified in the RREQ, are met. Once the route is established, if a node ceases to meet the QoS requirements for a specific flux, it sends an ICMP_ QOS_LOST message back to the source to inform it of this fact.

Two other mechanisms for guaranteeing QoS exist: maximum delay and minimum bandwidth for a route. To guarantee the delay, each time a node receives an RREQ, it subtracts the NODE_TRAVERSAL_TIME (the time taken for the message to reach the node) from the value transported by the RREQ.

To guarantee sufficient bandwidth availability, the value transported by the RREQ is compared with the capacity of the link. If the latter value is lower of the two, then the packet is rejected. When a destination replies to a source using an RREP message, each intermediary node compares the bandwidth field of the RREP with its own capacity and reinitializes this field with the minimum value before retransmitting the message.

This extension of the AODV protocol to support QoS has been the subject of a project where the author only introduced the delay metric.

The results obtained from the implementation of QoS_AODV in ns showed improvements in the level of packets successfully delivered (PDR), but an increase in control traffic. However, QoS_AODV seems to be less sensitive to network densification than its predecessor.

4.2.2.4. *MAC layer*

Multiple access collision avoidance with piggy back reservation (MACA/PR) is an extension of the IEEE 802.11 norm, which guarantees bandwidth using resource reservation for real-time traffic. MACA/PR is a medium access protocol, which allows real-time connections to a hop (or a link) to be established.

It allows flux differentiation at access level alongside rapid and reliable transmission of nonreal-time traffic. MACA/PR uses a reservation table, which aims to block the transmission of nonpriority traffic. Priority traffic prevents other fluxes from being sent along the same path as itself.

4.3. The OLSRQSUP protocol and QoS extensions

To establish a session, a routing protocol with QoS must find a route with the QoS constraints specified. For example, a route may be required with the maximum possible bandwidth. Demands of this kind are currently difficult to meet.

We are faced with an NP-complete problem. An NP-complete problem is a decision problem for which a very large number of cases must be examined. In these cases, a polynomial algorithm that would provide the solution does not always exist. In brief, if QoS contains at least two additive metrics, QoS routing becomes an NP-complete problem. To clarify the notion of additive and concave metrics, we shall use the following definitions:

Definition of Concave and Additive QoS Metrics

– Let $m(i,j)$ be a QoS metric for the link (i,j). For a path $P = (s,i,j, \ldots, l,t)$, we say that a metric m is concave if:

$$m(P)=min\{m(s,i),m(i,j), \ldots, m(l,t)\}.$$

The metric is additive if:

$$m(P)=m(s,i)+m(i,j)+\cdots+m(l,t).$$

Examples: Bandwidth is a concave metric, whereas delay is an additive metric.

Routing with QoS in ad hoc networks is extremely complex due to constraints imposed by the nature of the network itself, including mobility and variable topology, and by the limited bandwidth available to carry supplementary QoS routing control traffic. The maintenance of precise link-state information places a considerable burden on networks of this kind. Proactive protocols are less sensitive in this regard than on-demand protocols as the former maintain bases of information to complete the task.

The new protocol will introduce delay and bandwidth parameters into the various bases. Multipoint relay (MPR) searches will then be improved using these two metrics. Routing will be based on the same two metrics. Further details will be given in the following sections.

4.3.1. *Operation of the protocol*

To improve QoS in an ad hoc network, parameters such as delay, bandwidth, packet loss, and error levels should be taken into consideration. However, as we have already explained, the addition of multiple metrics would greatly complicate the problem to the point of becoming unsolvable. For this reason, we have only selected the two parameters most useful in assuring QoS.

4.3.1.1. *Delay*

The OLSRQSUP protocol includes the delay parameter for each entry associated with different tables. The basic principle is to measure the time delay between a node and its neighbor. However, measuring the delay is not a simple task, given the wireless environment. To arrive at an optimal approximative solution, we shall look at the case of synchronized networks. We shall assume that the mobile nodes are equipped with a global position system (GPS) system. Nodes in this network are synchronized regularly to ensure that all their clocks show the same time.

In our proposition, each HELLO message issued by a node includes the time of creation of the message. During the neighborhood discovery phase and when the HELLO message is received by the neighboring node, delay is calculated by comparing the time in the message with the clock time. This delay time obviously includes any queuing time, transmission time, and collision avoidance time. To give an example using the IEE 802.11 norm, the delay is calculated as follows:

$$\text{Measured delay} = t_q + (t_s + t_{\text{ca}} + t_{\text{overh}}) \times R + \sum_{R=1}^{R} B_T \quad [4.1]$$

where t_q is the waiting time (time queue), t_s the time the message was sent (sent time), t_{ca} the time taken for the collision detection phase (collision avoidance), t_{overh} the overhead control time (such as Request To Send (RTS), Carrier To Send (CTS), and Acknowledgement (ACK)), R the number of retransmissions needed, and B_T the backoff time.

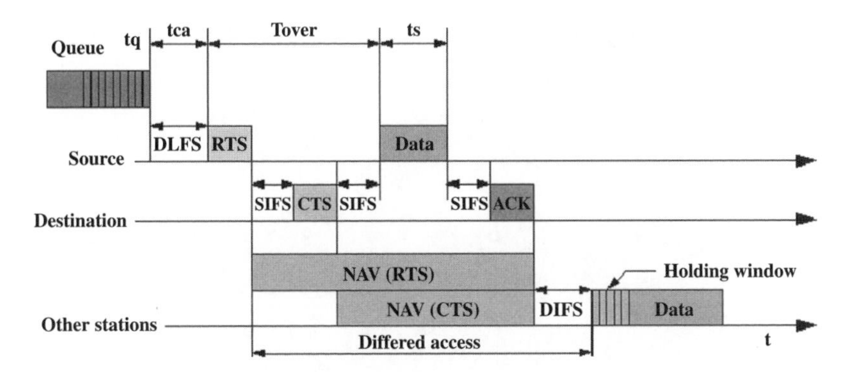

Figure 4.3. *Transmission framework*

To provide a higher level of precision, we will measure average delay, i.e. if the most recent value for the delay is 20 ms, whereas the precedent was 2 ms, then a value of 1 ms will be retained:

$$D_{ave} = \frac{D_{ant} + D_{actu}}{2} \quad [4.2]$$

where D_{ave} is the average delay, D_{ant} is the previous value contained in the table, and D_{act} the delay obtained upon reception of a HELLO message.

4.3.1.2. *Bandwidth*

The bandwidth metric can be defined as the bandwidth available to the link between the source and the destination. However, in a wireless environment, bandwidth can vary as a result of node mobility. A saturated link indicates that a neighboring node is transmitting. It may also indicate that a link within transmission range is using the channel, giving a high probability of collisions.

This metric will then be used in selecting MPRs and in routing calculations. As OLSR always selects routes using MPRs, we may presume that, added to the delay metric, the bandwidth metric will assist in the choice of an optimal route.

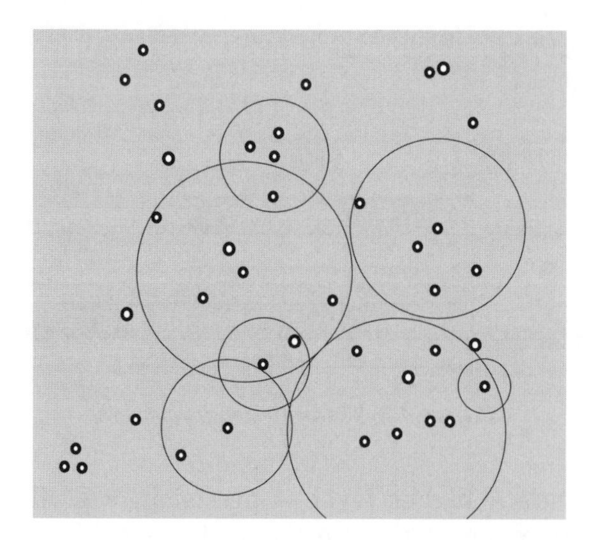

Figure 4.4. *Collisions in a wireless network*

Nevertheless, the estimation of this parameter is extremely complex due to the nature of ad hoc distribution and the nature of the link. Various publications have appeared on this subject.

Chen and Heinzelman, for example, have recently proposed a routing protocol with QoS based on the AODV protocol. In their article, the authors suggest two means of estimating available bandwidth.

One method consists of "listening" to the channel and calculating the bandwidth using the difference between idle time and busy time. However, although this method is precise, it does not take the effects of mobility into account.

The second method suggested involves the use of HELLO messages to determine bandwidth and to free up resources in case of route failures.

Our proposition for estimating bandwidth is simple and intuitive. We will presume that the terminals used in the ad hoc network are all configured with the same wireless card, i.e. the IEEE 802.11 norm. Thus all the nodes in the network have the same maximum bandwidth. We estimate the maximum debit at 2 Mbps. This bandwidth is not the most optimal, but the simplest to calculate as we are interested in what remains for the traffic.

The size of messages sent and received is recorded over a fixed period, T. If N is the number of packets received and S is the size in bits, then the average bandwidth use is given using equation [4.3]:

$$\mathrm{BW(bps)} = \frac{N \times S \times 8}{T} \qquad [4.3]$$

The level of precision with which bandwidth is calculated depends on the time parameter T between different measurements. The higher the value of T, the more precise the measurements will be. However, T must be small enough to deal effectively with mobility in the network. We are thus forced to compromise between transparency and precision.

Using this method, every packet that crosses the transmission channel and every packet passing within the carrier sensing range of the mobile node will be taken into account in bandwidth calculations. To simplify the routing algorithm including the available bandwidth metric, we intend to check the routing table with each bandwidth calculation.

4.3.2. *Sensing of neighborhood QoS parameters*

Each network node must detect neighbors with which it has a direct symmetric link. The radio link, by its nature, is prone to interruptions. For this reason, the HELLO messages sent out by nodes within their neighborhood contain link-state information. Each node must estimate the QoS parameters (bandwidth, delay, etc.) and those of its direct neighbors with which it shares a bidirectional link. QoS parameters are then diffused using the appropriate control messages.

4.3.2.1. *HELLO message extensions*

The HELLO message extensions supplied aim to allow neighboring nodes to detect various QoS parameters governing the network.

0	1	2	3
0 1 2 3 4 5 6 7	8 9 0 1 2 3 4 5	6 7 8 9 0 1 2 3	4 5 6 7 8 9 0 1
Reserved		**Htime**	**Willingness**
Link Code	**Reserved**	**Link Message Size**	
Neighbor Interface Address			
Delay		**Available Bandwidth**	
Neighbor Interface Address			
Delay		**Available Bandwidth**	
……..			
Link Code	**Reserved**	**Link Message Size**	
Neighbor interface Address			
Delay		**Available Bandwidth**	

Figure 4.5. *HELLO message format for OLSRQSUP*

The HELLO message presented above contains the following:

– Address list of symmetric neighbors with values of QoS parameters over the respective links.

– Address list of heard neighbors with respective QoS parameter values.

– MPR list with QoS values for links that connect the local node to its neighbors.

The *Reserved*, *Htime*, *Willingness*, *Link Code*, *Link Message Size,* and *Neighbor Address* fields are identical to those used in OLSR and have already been described.

QoS parameters are given for each neighbor using:

– *delay*: (16 bits) – delay measurement given in milliseconds and

– *available bandwidth*: (16 bits) – bandwidth measurement for the link in question, given in bits per second (bps).

4.3.2.2. *Format of information base extensions*

To support the new metrics introduced into the protocol, an extension for the information bases used is also required.

To do this, we inserted delay and available bandwidth fields into each field, transmitted using HELLO messages.

In other words, nodes accumulate values obtained in tables, presented as follows.

4.3.2.2.1. Neighbor set

N_addr	N_status	N_willingness	N_bandwidth	N_delay	N_time

Table 4.1. *Format of neighbor set table extension*

In the above table, N_addr is the address of the neighbor, N_STATUS the state of the neighbor (i.e. MPR or symmetric), and N_willingness an integer from 0 to 7 inclusive indicating the capacity of the neighbor to transmit traffic for other nodes. This parameter is important later in the choice of MPRs. N_bandwidth is the bandwidth for the link between the node and its neighbor, given in bps. N_delay is the mean time taken to transmit a HELLO packet between the two neighbors, given in ms. N_time is the time for which the information given should be considered valid and after which it should be deleted from the table.

4.3.2.2.2. Two-hop Neighbor Set

N_addr	N_2hop_addr	N_2hop_bandwidth	N_2hop_delay	N_time

Table 4.2. *Format of two-hop neighbor set table extension*

N_addr is the address of the neighboring node, which allows contact with the node N_2hop_addr with a delay of N_2hop_delay from the local node and available bandwidth estimated at N_2hop_bandwidth. This information becomes invalid after N_time.

4.3.2.2.3. MPR selector set

MS_addr	MS_bandwidth	MS_delay	MS_time

Table 4.3. *Format of MPR neighbor set table extension*

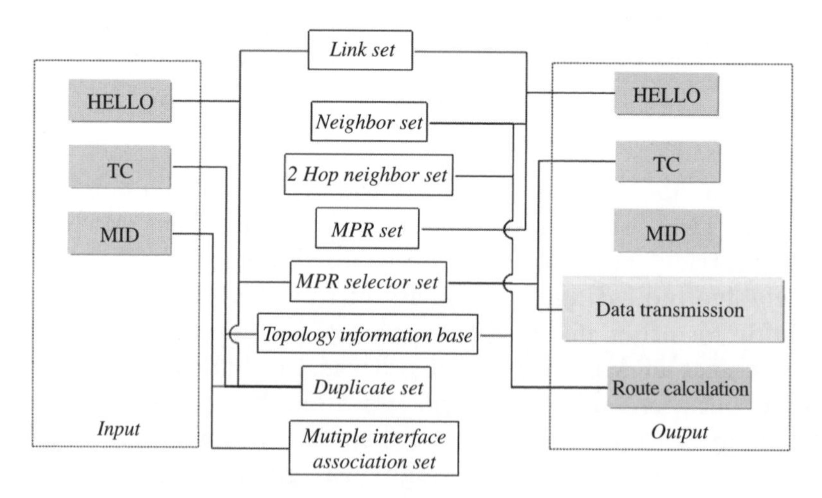

Figure 4.6. *Global architecture of the protocol*

Table 4.3 gives the address of the node MS_addr which selected the local node as a relay point for its traffic, with a delay of MS_delay between the two and available bandwidth of MS_Bandwidth over the link connecting the nodes.

Figure 4.6 gives a global overview of the operation of the protocol showing the functions of different tables.

4.3.2.3. *MPR selection algorithm*

The MPR selection algorithm does not take delay and bandwidth parameters into account. Links with high bandwidth and minimal delay may, therefore, not be selected as MPRs. Clearly, a new MPR selection algorithm is required which will take these parameters into account.

This algorithm is used by each node to optimize selection within the norms of QoS.

Before presenting the new algorithm, we need to clarify the notations used in Figure 4.6:

– $N(x)$: all of the neighbors of node x at a distance of one hop;

– $N2(x)$: all two-hop neighbors of x;

– $R(y)$: for each node, y being part of $N(x)$, we can calculate R, defined by the number of nodes in $N2(x)$ not covered by at least one neighbor;

– $Del(b)$: the delay between the node in question and the local node; and

– $B(y)$: the respective available bandwidth.

The MPR selection algorithm is presented in Figure 4.7.

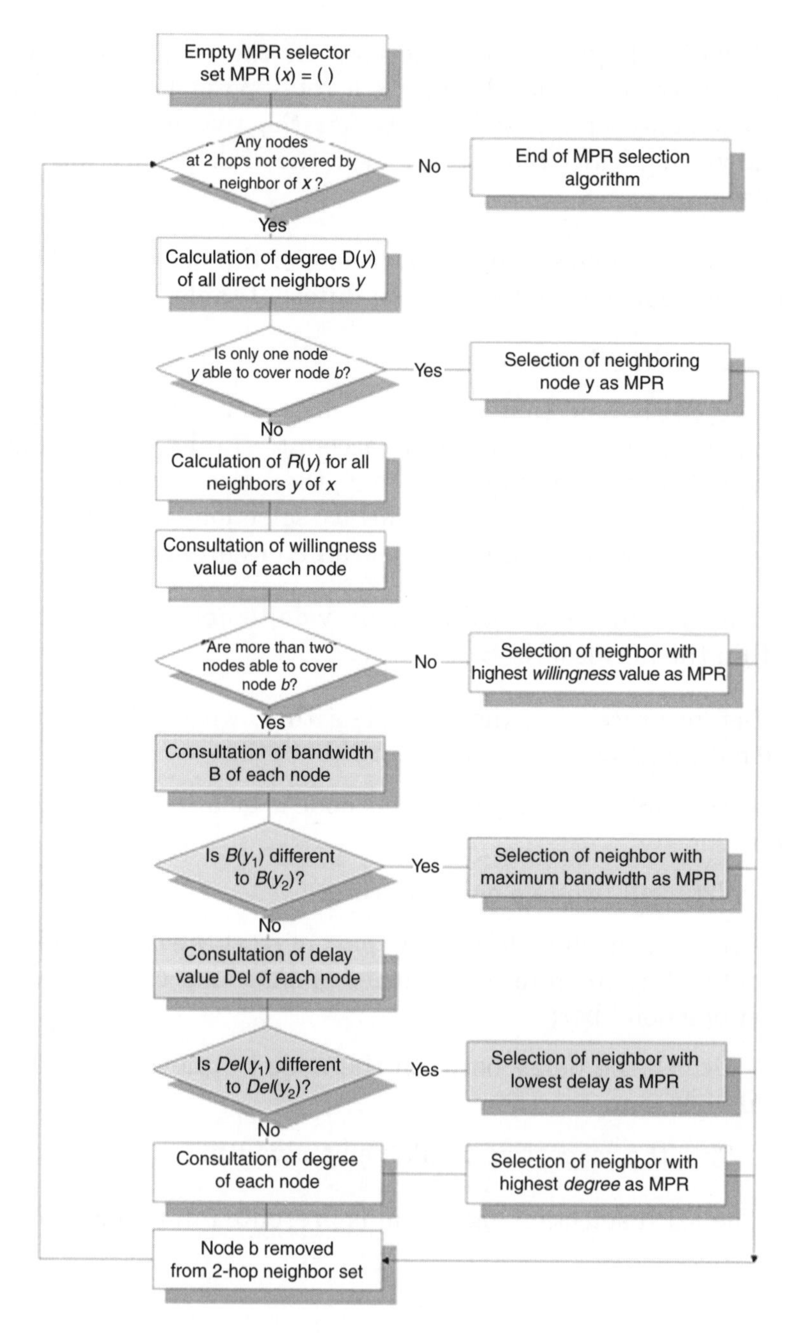

Figure 4.7. *Multipoint relay selection algorithm*

An example explaining this algorithm is shown in Figure 4.8.

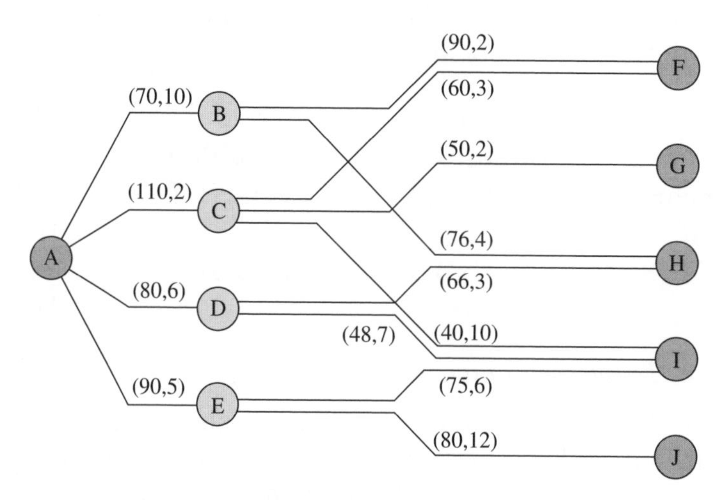

Figure 4.8. *Example execution of MPR selection algorithm for the OLSRQSUP protocol*

When selecting relay points, node A will consider its neighbors at one hop which allow it to transmit traffic to nodes further away.

To do this, OLSR concentrates on those nodes that allow it to reach several more distant neighbors, in this case *E*, *C*, and *B*.

Node *E* offers a link to node *J* and is the only node offering this connection.

However, OLSRQSUP concentrates on those neighbors offering more bandwidth and lower delay. Node *D* is therefore selected as a relay point instead of node *B* for the following reasons:

– The link from *A* to *H* through node *B* has a weight of (70.10) + (76.4) = (146.14), where the first parameter is bandwidth and the second the delay metric.

– The path to *H* via node *D* has a weight of (80.6) + (66.3) = (146.9). This path therefore offers the same bandwidth but lower delay.

	1 hop neighbor	**2 hop neighbor**	**MPR**
Standard OLSR	B, C, D, E	F, G, H, I	E, C, B
OLSRQSUP	B, C, D, E	F, G, H, I	E, C, D

4.3.2.4. *Topology management*

The OLSRQSUP protocol operates in the same manner as OLSR, but with an extended topology control (TC) message format. TC messages take the form shown in Figure 4.9.

0 1 2 3 4 5 6 7 8 9 0 1 2 3 4 5	6 7 8 9 0 1 2 3 4 5 6 7 8 9 0 1
ANSN	Reserved
Advertised Neighbor Main Address	
Delay	Available Bandwidth
Advertised Neighbor Main Address	
Delay	Available Bandwidth
Advertised Neighbor Main Address	
Delay	Available Bandwidth
...	

Bit offsets: 0, 1, 2, 3

Figure 4.9. *TC message format in OLSRQSUP*

The *Delay* and *Available Bandwidth* fields have been added to enable QoS parameters to be saved in tables and diffused using the TC message. Both fields have a size of 16 bits. Delay is measured in milliseconds and debit in bps.

The network topology table takes the format shown in Table 4.4.

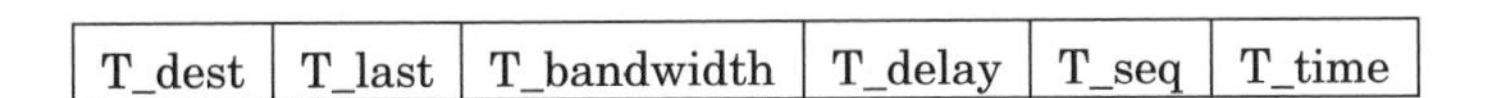

T_dest	T_last	T_bandwidth	T_delay	T_seq	T_time

Table 4.4. *Format of topology table with extensions*

For each destination in the network, the topology table contains the address of the node, T_dest. The node may be reached through the neighbor T_last with a delay of T_delay. T_seq and T_time are the sequence number and validity time of the tuple, respectively.

4.3.2.5. *Routing*

Each network node maintains a routing table with the format shown in Table 4.5.

R_addr	R_next	R_dist	R_bandwidth	R_delay

Table 4.5. *Routing table format*

To fill out the routing table, the node no longer considers the hop count metric, but the available bandwidth and delay. In other words, the node no longer seeks the shortest path in terms of the number of hops, but the optimal route in terms of QoS.

4.4. Conclusion

In this chapter, we have presented an overview of research already carried out with the aim of providing QoS in an ad hoc environment. One of the subjects discussed in this chapter was QoS-oriented routing. This domain is particularly interesting in that most ad hoc routing protocols do not provide optimal routing in terms of QoS even after standardization.

For this reason, we shall deal with this theme in the present work to introduce extensions to improve the OLSR protocol. We have already presented various extensions to OLSR that allow it to support QoS, and the MPR selection algorithm has been modified to respond to this need. Routing is then carried out using two metrics, delay and bandwidth. In Chapter 5, we shall present the steps of implementation of the OLSRQSUP protocol in ns and the simulations carried out to evaluate the performance of the new protocol.

Chapter 5

Implementation and Simulation

5.1. Introduction

Every protocol design, with or without quality of service (QoS), needs to be backed up by a performance evaluation. The use of real networks for this, particularly in the case of ad hoc networks, is difficult and expensive. Moreover, evaluations of this kind do not give useful results. A real network does not offer adjustable and easily controllable environmental parameters and the extraction of data is problematic.

For this reason, most performance evaluations and research in this domain are based on simulations. Simulations not only make it possible to test new technologies and new protocols while varying different parameters but also allow us to anticipate future results and potential problems with the protocol.

In this chapter, we provide a detailed description of the implementation of the OLSRQSUP protocol using the ns simulator. Next, we define the context of the simulation. Finally, we discuss the results of this simulation and offer an interpretation of the phenomena observed.

5.2. Implementation

5.2.1. *Use of the simulator*

Before explaining the various stages of programming involved in this study, we must clarify certain points, including the background knowledge required for the different steps. The ns simulator remains the most widespread program of its kind for research purposes.

Almost all studies already published in the domain of ad hoc networks were carried out using the ns simulator, despite a recent tendency to use OPNET. This tendency is due to the complexity of ns, which has pushed some researchers to use the simpler Modeler version of OPNET.

To use the ns-2 simulator, a certain level of background knowledge is required:

– Knowledge of the C++ and familiarity with concepts of object-oriented programming to edit the routines of the simulator.

– Knowledge of the OTCL script language for description of simulations and pretreatment of data.

– Familiarity with the AWK, the tool used for data extraction.

– Ability to produce graphs from the files produced by the simulation, using Gnuplot, Xgraph, or Matlab.

There are several steps involved in a simulation:

– Reading the script, which specifies the protocol being used and the parameters of the simulation (topology, duration, and scenarios).

– Extraction of the protocol operating mechanism by the simulator using the routines.

– Production of nam and trace files.

Figure 5.1 shows the various steps of a simulation in ns-2.

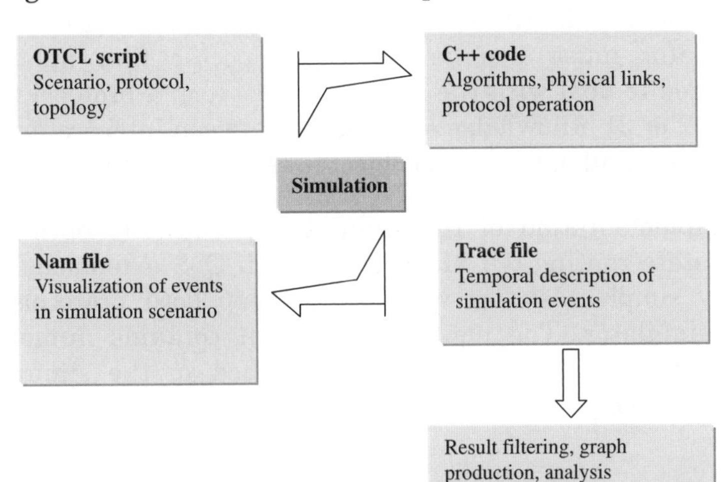

Figure 5.1. *Stages of ns-2 simulation*

The simulator can be used on different levels depending on the experience of the user, from novice to expert:

– Basic usage allows the user to simulate ns as it is, i.e. the user is able to modify a tcl script to modify simulation scenarios. At this level of usage, parameters can also be modified; the user may change the value of a given parameter, for example data refreshment time, in the tcl script. The simulator will then use the value contained in the script instead of the default value as defined in /tcl/lib/ns-default.tcl.

– Intermediate use of the simulator involves modifying the code (in C++), but without modifying the core of the simulator. For example, the intermediate user might modify a routing algorithm. In this case, the user would then need to recompile ns to check the syntax and generate the .o files necessary for the operation of the protocol.

– Advanced use of ns allows an expert user to develop their own code (in C++). In this case, the core of the simulator must also be modified. Specific files must be changed to deal with the added protocol (e.g. a protocol from layer 2 or 3). Knowledge of the links between interpreter and simulator and of the model object is essential in this case.

Implementation of the simulator to evaluate optimized link-state routing (OLSR) protocol with QoS extensions is a fairly simple task, as we wish to keep both protocols on the simulator. The new routing agent contains numerous properties of the "agent" class supplied in the simulator (Figure 5.2).

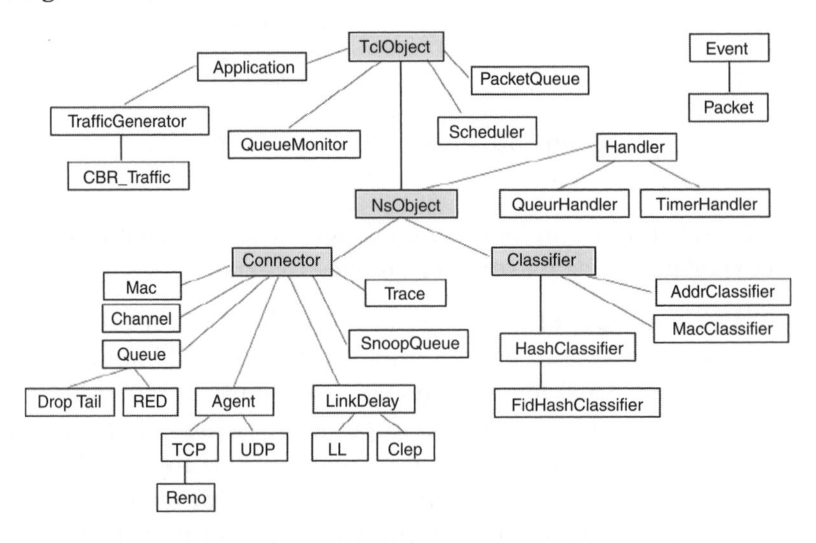

Figure 5.2. *Derivation tree of C++ classes of the simulator*

5.2.2. *Implementation by steps*

In implementing the protocol, we chose a strategy that does not necessitate modification of the simulator. We instead modified the mother-agent to check for syntax errors. We then implemented the OLSRQSUP protocol to generate scripts (Figure 5.3).

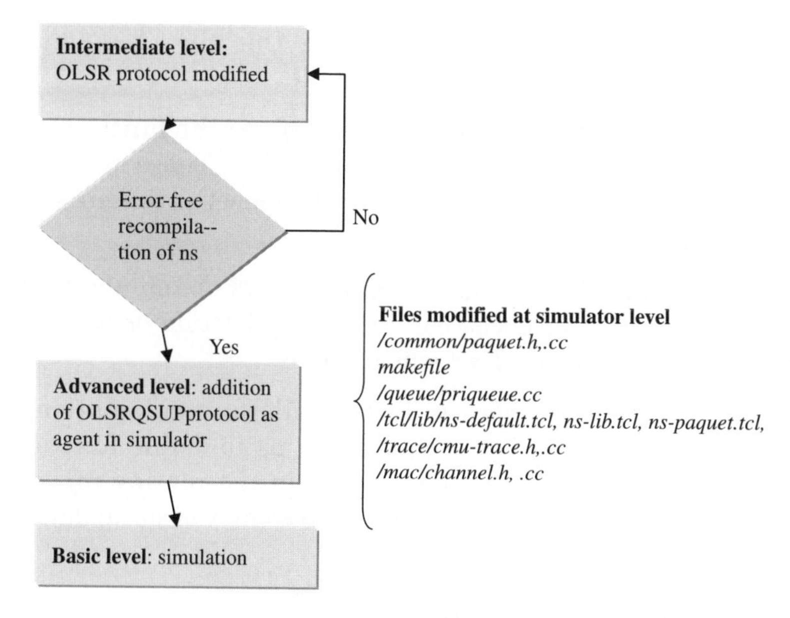

Figure 5.3. *Stages of implementation of the OLSRQSUP protocol at ns-2 simulator level*

5.2.3. *OLSRQSUP modules*

The code developed in the course of this project consists essentially of a group of C++ files:

– *OLSRQSUP_packet.h*. The file containing the format of the various control messages used by the protocol (HELLO, topology control (TC), and MID).

– *OLSRQSUP_repositories.h*. Section of code defining the format of the table entries that govern the operation of the protocol.

– *OLSRQSUP (.h, .cc)*: *main section, or skeleton, of the protocol*. The routing agent proper is defined in this file, which includes the routing and multipoint relay (MPR) selection routines. The procedures used to manage various events in nodes, for example the reception of HELLO or

TC messages, are also contained in this file. Temporizer management is also specified in the file.

– *OLSRQSUP_printer (.h, .cc)*. It defines the functions for the display of various information tables. These functions can be called up using the .tcl script to show the contents of a given table in the trace file.

– *OLSRQSUP_state (.h, .cc)*. It allows manipulation of tables within the protocol, e.g. insertion, deletion or updating of a tuple, or removal of the whole table.

– *OLSRQSUP_rtable (.h, .cc)*. It allows manipulation of node routing tables. These files allow us to define functions in routing tables and are very useful in the routing process, for example when seeking a specific table entry or for updating.

5.2.4. *Calculation of metrics*

To calculate the delay metric each time a HELLO message is received, we presume that all nodes in the network have synchronized clocks. However, the connectivity management processes of the protocol require some modifications. The time of creation (the timestamp) must be included in a field of the packet. Any node receiving a HELLO message can then calculate the propagation delay between itself and the neighboring node. This information is then stored in the neighbor information table. A typical table entry of this kind is given in Chapter 4.

In parallel to the activity described above, the node "listens" to the radio channel to determine the value of the bandwidth over the link between itself and the node that sent the HELLO message. Each node maintains a cache containing the available bandwidth with the group of nodes within transmission range. This cache is updated every 0.5 s. A typical treatment is given in the following list.

```
/// We need to know the delay for each received packet and BW
/// Delay's unit is double as the timestamp(milliseconds) and BW(bps)
double del = Scheduler::instance().clock();
double delay = (del - ch->timestamp())*1000;
double avail_bw = WirelessChannel::instance().availBW(index);
/// for QoS management
```

The cumulative delay will then be calculated. In other words, to fill in the table for two-hop neighbors, then end-to-end delay is calculated. A similar treatment is then applied to the bandwidth, allowing neighbor tables to be updated.

The calculated delay and estimated bandwidth are used in the calculation of the full set of tables. Each time a HELLO message is received, the following operations are carried out:

```
link_sensing(msg, receiver_iface, sender_iface,
delay_sen_rec, bw_sen_rec);
populate_nbset(msg, delay_sen_rec, bw_sen_rec);
populate_nb2hopset(msg, delay_sen_rec, bw_sen_rec);
mpr_computation();
populate_mprselset(msg, delay_sen_rec, bw_sen_rec);
```

In this way, link detection and updates to the neighbor set and two-hop neighbor set are carried out, alongside the execution of the MPR selection algorithm and calculation of the table of those neighbors using the node as MPR. OLSRQSUP inherits these processes from its predecessor, OLSR.

5.3. Simulation

5.3.1. *Simulation parameters*

The general parameters used for the simulation are summarized in Table 5.1.

Simulation context parameters	
Simulation time	180–300 s
Number of scenarios per context	5
Protocol	OLSR/OLSRQSUP
Number of nodes	30–50
Surface occupied by network (*grid / strip*)	1,000 × 1,000 m– 2,000 × 500 m
Maximum number of connections	5–30
Size of data packets	64 bytes
Send velocity	4 packets per second
Size of node queues	50 packets
Maximum node velocity	0–20 m/s
Maximum pause time	0 s

Table 5.1. *Simulation parameters*

5.3.2. Parameters to evaluate

The parameters to evaluate for the simulations are as follows:

– packet delivery ratio (PDR) (equation [3.4]);

– average throughput received (ATHR), used to calculate the useful debit for the whole of a given flux from sending to completion of reception (this parameter is defined as:

$$\text{Average throughput received (ATHR)} = \frac{\sum_{1}^{F} \frac{N \times S}{\text{TR}_N - \text{TE}_1}}{F} \text{ (Kbps)} \qquad [5.1]$$

where F is the number of fluxes generated in the network, N the number of packets generated per flux, S the size of a packet, TR_N the time of reception of the last packet, and TE_1 the time the first packet of the flux was sent. This parameter gives the capacity of the network);

– the volume of control traffic or traffic overhead (TOH) (equation [3.6]);

– the average end-to-end delay (equation [3.3])

– the average length of a path, defined as follows:

$$\text{Average packet length}(APL) = \frac{\sum_i S(i)}{\sum_j \text{packet received}} \quad \text{(hops)} \qquad [5.2]$$

where $S(i)$ is the number of hops made by the packet i to reach its destination;

– the average MPR number (this parameter gives the average number of MPRs in the network. It is calculated every 5 s (Appendix 3).

5.3.3. *Simulation results*

5.3.3.1. *Quantitative results*

We first evaluated various parameters according to velocity. Table 5.2 gives the results of simulations for two networks of different density (30 and 50 N/km^2) with variable velocity. The number of fluxes is fixed at five connections with a maximum emission of 1,000 packets of 64 bytes each. The duration of the simulation is fixed at 300 s.

Figures 5.4 and 5.5 are .nam file captures resulting from the simulation.

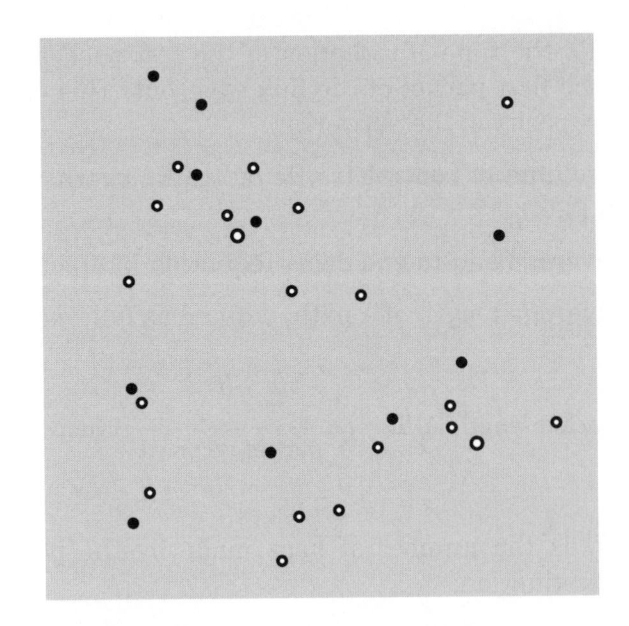

Figure 5.4. *1,000 × 1,000 m network with 30 nodes*

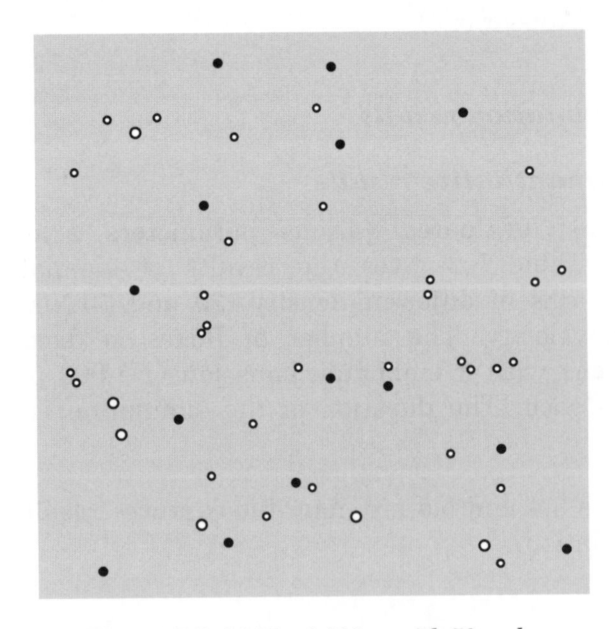

Figure 5.5. *1,000 × 1,000 m with 50 nodes*

Second, we evaluated the impact of topology on a network using the topology shown in Figure 5.6.

Each value in Table 5.3 is the result of five simulations of different scenarios for the same velocity, each lasting 180 s. In this way, for the whole simulation the computer takes 75 min.

In fact, the whole process takes a longer time; the time needed to process the script and generate the trace file is greater than that required for the actual simulation, which is the time during which the network is communicating. The total duration of the process explains our choice of a relatively short simulation time and the use of four flux values only.

Packet size is fixed at 64 bytes, and packets are sent at a rate of four per second. Each flux generates a maximum of 1,000 packets. The network contains 50 nodes distributed in a linear architecture (i.e. a strip) rather than the grid used previously. This choice stems from the fact that this type of network can be used in a reduced mobility situation on university campuses (Figure 5.6).

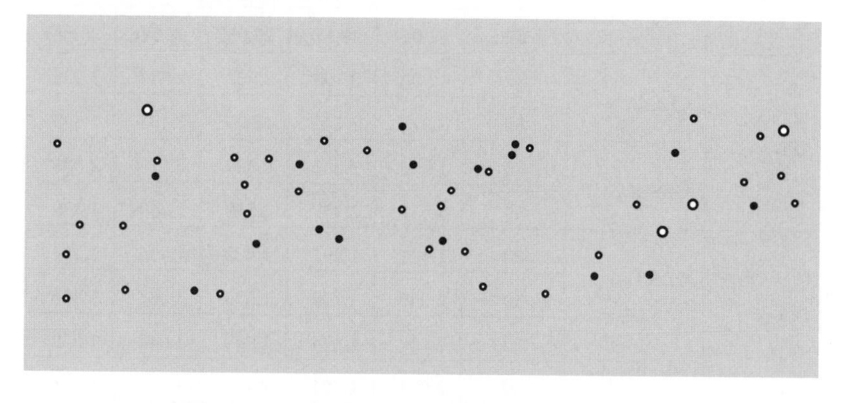

Figure 5.6. *Topology of 50 nodes in strip layout*

Parameter	Protocol	Velocity Number of nodes	0	5	10	15	20
PDR	OLSR	30	89.839	83.062	77.502	75.193	72.751
		50	98.451	89.541	85.845	79.547	75.638
	OLSRQSUP	30	87.983	72.742	69.818	64.197	61.508
		50	94.879	84.412	81.61	76.699	67.801
APD	OLSR	30	6.706	8.891	13.414	19.785	27.226
		50	9.897	10.352	22.542	28.490	37.907
	OLSRQSUP	30	6.926	7.992	10.254	13.785	19.226
		50	5.100	5.352	6.542	7.388	8.020
MPR	OLSR	30	13.946	21.172	22.319	22.534	22.924
		50	28.457	35.352	37.323	38.588	41.352
	OLSRQSUP	30	16.376	22.206	24.81	24.913	25.293
		50	43.512	44.082	44.441	46.285	46.6
TOH	OLSR	30	11,011	12,153	12,320	12,488	12,577
		50	14,271	16,003	16,333	16,666	19,968
	OLSRQSUP	30	12,890	12,621	13,245	13,980	13,836
		50	21,974	23,126	23,587	23,788	25,885
APL	OLSR	30	1.922	2.108	2.057	2.537	2.339
		50	3.317	2.146	1.997	2.493	1.695
	OLSRQSUP	30	1.922	2.057	2.537	2.17	2.339
		50	3.536	2.257	2.519	2.259	1.369
ATHR	OLSR	30	1.601	1.591	1.629	1.418	1.174
		50	2.056	1.772	1.505	1.561	1.356
	OLSRQSUP	30	1.663	1.874	1.657	1.342	1.259
		50	2.081	1.921	1.764	1.616	1.444

Table 5.2. *Simulation results: impact of mobility and table density*

Protocol	Number of fluxes	PDR (%)	APD (ms)	TOH (pkts)	MPR (nodes)	APL (hops)	THR (Kbps)
OLSR	5	99.851	9.106	12,271	32.514	3.317	1.712
	10	98.925	10.188	12,900	30.647	3.435	2.008
	15	98.796	17.506	13,000	29.348	4.566	2.012
	20	96.276	25.978	12,739	31.205	3.552	2.051
	25	92.487	26.852	11,852	26.254	3.003	1.813
	30	84.339	30.967	12,967	24.587	5.064	1.610
OLSRQSUP	5	99.456	6.249	18,312	42.029	4.318	1.949
	10	98.781	8.586	19,303	42.705	4.270	2.146
	15	96.514	14.127	18,794	42.000	5.077	2.169
	20	92.926	19.356	19,594	40.218	4.204	2.279
	25	86.754	20.043	18,817	41.256	4.004	2.007
	30	83.625	25.417	18,938	40.875	3.562	1.810

Table 5.3. *Simulation results: impact of number of connections*

5.3.3.2. *Impact of mobility and network density*

5.3.3.2.1. Packet delivery ratio

Figure 5.7 shows the PDR for a 1,000 × 1,000 m network, using four types of density for each protocol with different velocities. The reason studying this parameter first is fairly obvious, as it covers the main function of a protocol, i.e. the transmission of traffic.

It should be noted that, for the four scenarios represented in Figure 5.7, PDR diminishes as velocity increases. With reduced mobility, links between nodes are less likely to break; paths stored in the routing tables are thus more likely to be up-to-date, leading to a satisfactory PDR. In a static network, the OLSR protocol produces a PDR value of over 90% for different densities. This value is sufficiently high for use by applications with no retransmission mechanism. The OLSRQSUP protocol, however, is less successful in this respect, with a lower PDR value of 84%.

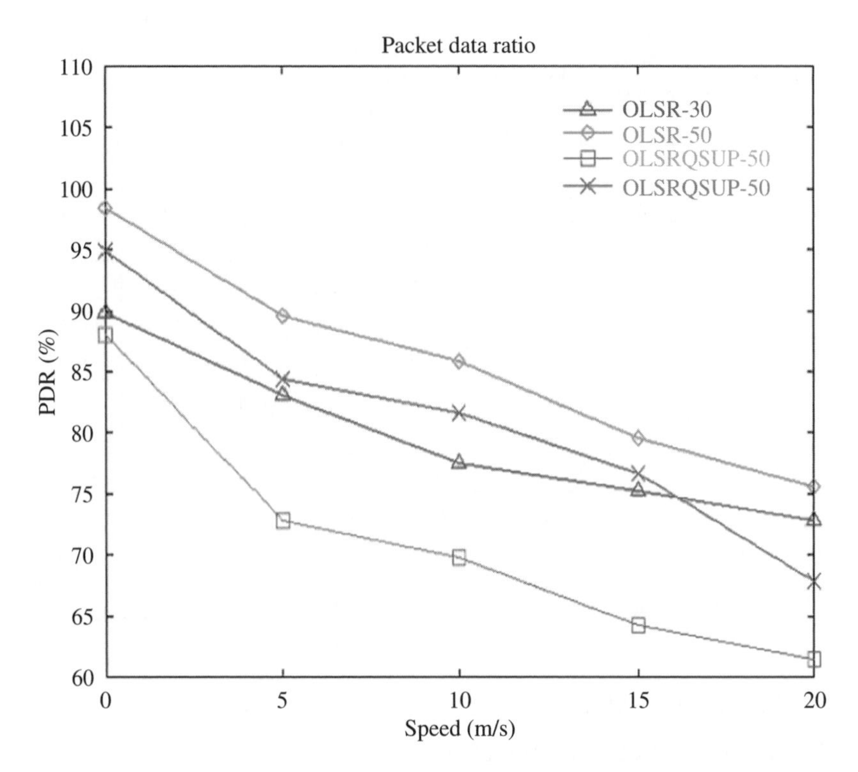

Figure 5.7. *PDR of a 1,000 × 1,000 m network depending on velocity*

The OLSRQSUP protocol consequently produces a large quantity of control traffic. With the OLSR protocol, a main aim was to reduce the TOH by reducing the MPR group. However, with the OLSRQSUP protocol, additional information on QoS needs to be shared with the network to provide routing with maximum bandwidth and minimum delay.

The number of MPRs is thus greater for the OLSRQSUP protocol than for OLSR (Figure 5.8). The TOH is consequently higher, reducing the availability of links and increasing the risk of message collisions. This leads to a degradation of performance and inefficient updating of the routing table.

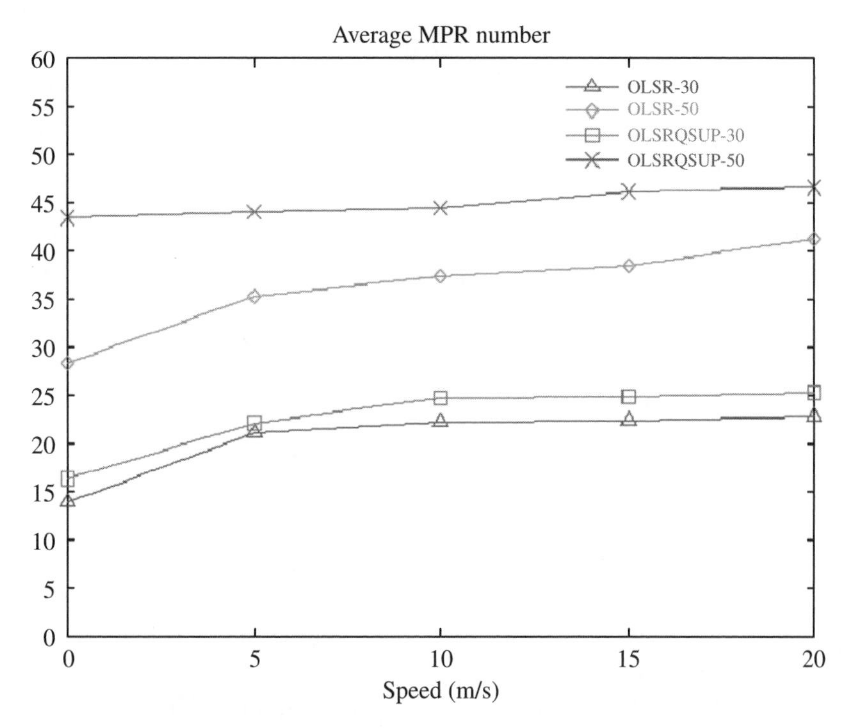

Figure 5.8. *Average MPR number in a 1,000 × 1,000 m network*

In Figure 5.9, based on the OLSR protocol, node 2 is selected as MPR by node 1, and generates a TC message to inform node 6 that a link exists between nodes 1 and 2. Nodes 3 and 4 are both MPRs for node 2, so diffuse the TC message.

Assuming that the medium is available at the instant in question, nodes 3 and 4 will send TC messages immediately, most probably at the same time. As a direct consequence, the TC messages will collide in node 6, which still does not know that a route to node 1 exists through node 2. According to the routing table of node 6, no route to node 1 exists.

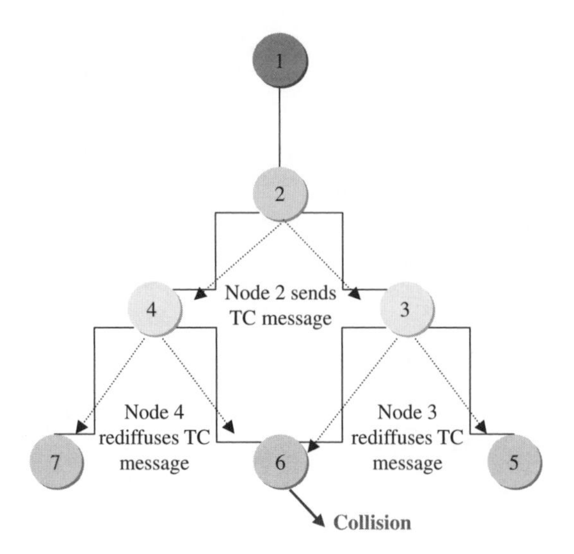

Figure 5.9. *TC message collision*

Through this example, we can see that if there is an overlap between two two-hop neighbors covered by several MPRs, there is an increased risk of collision. This effect is accentuated as mobility increases. For this reason, PDR is even lower at velocity, $V = 20$ m/s (70% for OLSRQSUP).

We also notice that densification increases PDR. For both the OLSR and the OLSRQSUP, a network of 50 nodes offers a better PDR than a network of 30 nodes. Intuitively, one would predict a lowering of performance with increased density. In actual fact, in a network with higher density, more routes to each destination are available.

The PDR reached by the OLSRQSUP is generally lower than that of the OLSR by around 12%, independent of velocity. This tells us that the OLSR and the OLSRQSUP have the same sensitivity to node velocity.

Another factor that plays a role in packet loss is queue size at node level. Queue size is limited to 50 packets, while nodes produce a higher quantity of outgoing traffic (1,000 packets). However, if queue size were increased, the benefits would be offset by increased delay.

So, in addition to the saturation of the network by control messages (Figure 5.10), several packets may wish to reach the same node and the capacity of the queue may be exceeded. OLSRQSUP aims to avoid this problem, as traffic is sent via the least saturated nodes. However, the effects of this improvement are not noticeable using the PDR parameter, as the TOH and the number of MPRs are both increased.

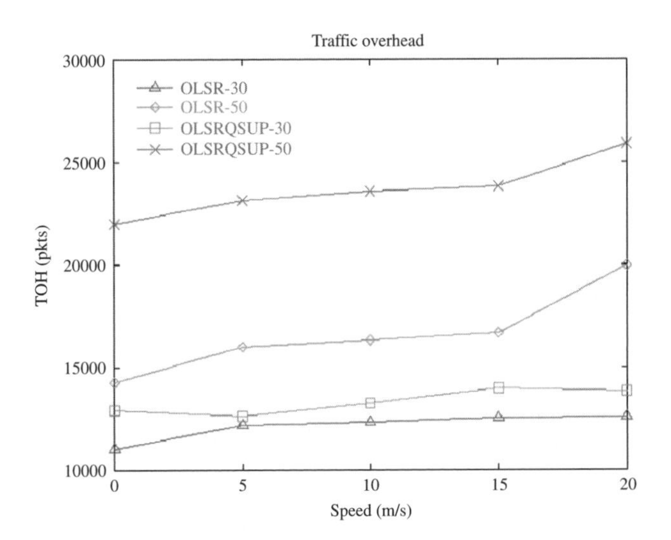

Figure 5.10. *TOH for a 1,000 × 1,000 m network*

5.3.3.2.2. Average packet delay (APD)

Figure 5.11 shows the average delay for transmission of a data packet from the source node to the destination node. This delay, calculated from end-to-end, only concerns those packets successfully delivered to their destination. Packets lost along the path following a route failure, whatever the cause, are not taken into account when calculating this parameter.

The aim of the OLSRQSUP protocol, which introduces QoS extensions, is to allow applications to benefit from optimized routing in terms of bandwidth and delay.

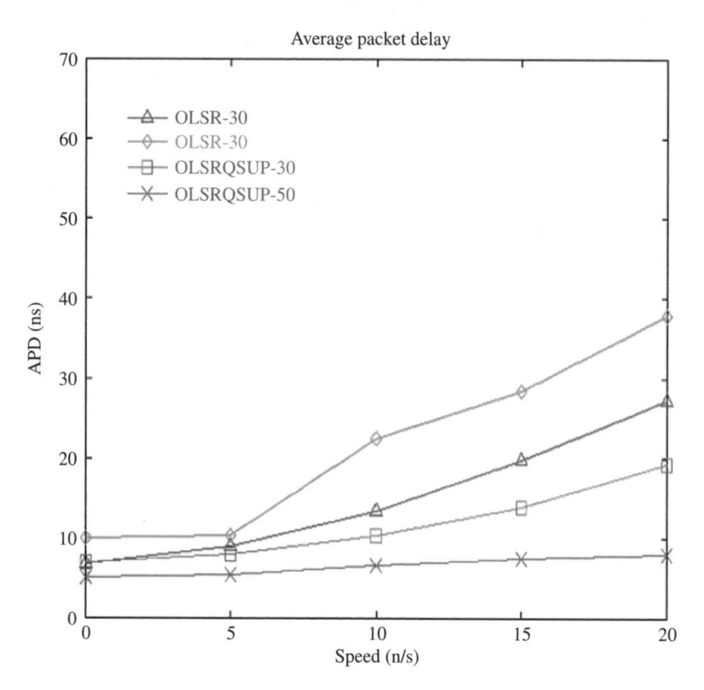

Figure 5.11. *APD for a 1,000 × 1,000 m network*

From Figure 5.11, we observe that delay in the OLSRQSUP is lower than that observed for the OLSR protocol in its best-effort routing form.

In the case of a static network (where velocity = 0), the transmission delay for a packet routed using OLSR in its best-effort form is 8 and 10 ms for networks with density of 30 and 50 N/km^2, respectively. The OLSRQSUP, with QoS extensions, transmits packets with a delay of 6 and 10 ms, respectively. This is due to the fact that the OLSRQSUP only considers paths with the two criteria already mentioned (maximum bandwidth and minimum delay). Consequently, as the network is static, the routes established avoid dense zones.

For different node velocities, the OLSRQSUP maintains an average difference of around 15 ms with its ancestor OLSR for a network of 50 nodes, and 7 ms for a network of 30 nodes.

This difference clearly demonstrates the contribution of the new routing protocol with QoS, and is due to the fact that a route from a source to a destination is only considered valid by OLSRQSUP if it fulfils the widest–shortest path criteria (i.e. path with highest bandwidth and lowest delay). The difference between results in networks of different sizes is due to the fact that the more nodes there are in the network, the longer the delays in node queues, and the more nodes the packet must pass through.

The OLSRQSUP protocol, however, is not affected in this way by the densification of the network, achieving transmission with a lower delay to that observed in a 30-node network. This can be explained by the fact that the OLSRQSUP sends traffic toward those nodes least subject to control traffic, presenting a "free" transmission zone. This reduces both the risk of collisions and the risk of packets having to wait to access a node.

For both proactive protocols, delay increases slightly with node mobility. This is due to frequent changes in network topology, meaning that a data packet may spend slightly longer in the buffers of the source node.

5.3.3.2.3. Average throughput received

This measurement represents the average useful debit of paths used by data packets. The evaluation of this parameter is important in that it shows the contributions made by our work. The modifications we made to the MPR selection algorithm effectively improve debit along the whole route.

In the first instance, this parameter appears to be very low (~1.8 kbps). It should be remembered that five fluxes of maximum 1,000 packets are generated at a debit of four packets per second, with a size of 64 bytes per packet, so: $64 \times 8 \times 4 = 2.048$ kbps.

It would be naïve to assume that this level of performance could be reached for the whole network (necessitating a PDR of 100%, absence of interference and collisions, and the availability of the entire bandwidth for data traffic).

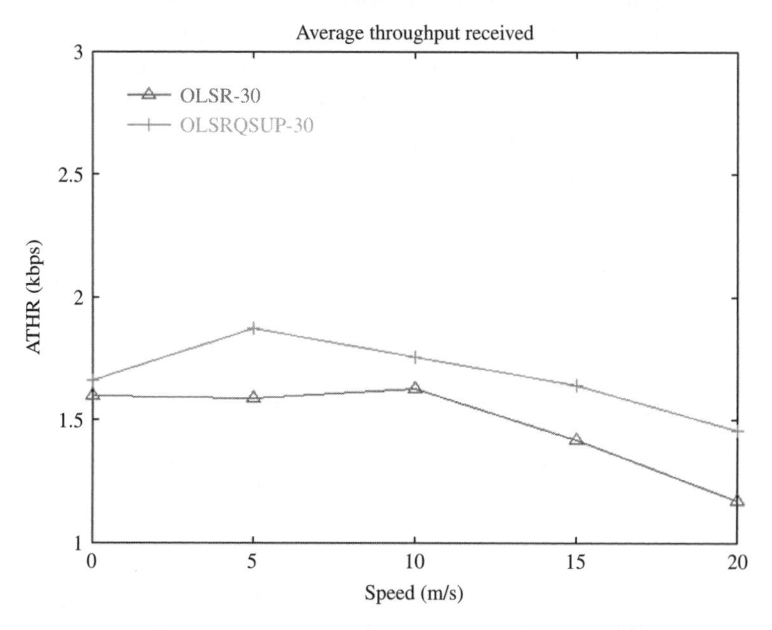

Figure 5.12. *ATHR for a 1,000 × 1,000 m network (30 nodes)*

Figure 5.12 shows that this parameter dependent on velocity for a network of 30 nodes. We notice a net improvement in ATHR, of around 6% for a static network and around 16% for a network with node mobility of 20 m/s.

At first glance, this seems unusual as OLSRQSUP has a lower PDR than its predecessor. In fact, this parameter is calculated using the average throughput of different fluxes. The fact that the proposed routes have more available bandwidth and less delay explains the improvement.

The OLSRQSUP protocol effectively seeks routes that maximize debit, i.e. routes with least congestion. It is interesting to note that the OLSRQSUP protocol allows traffic to be dispersed rather than concentrating it in a fixed zone. This allows route capacity to be increased, if only by 0.3 kbps.

It should be noted that this parameter could be considerably improved in a much larger network. Simulations carried out using OLSR in the past have tended to use a much more powerful simulator, due to the fact that this protocol requires high capacity memory and processing power.

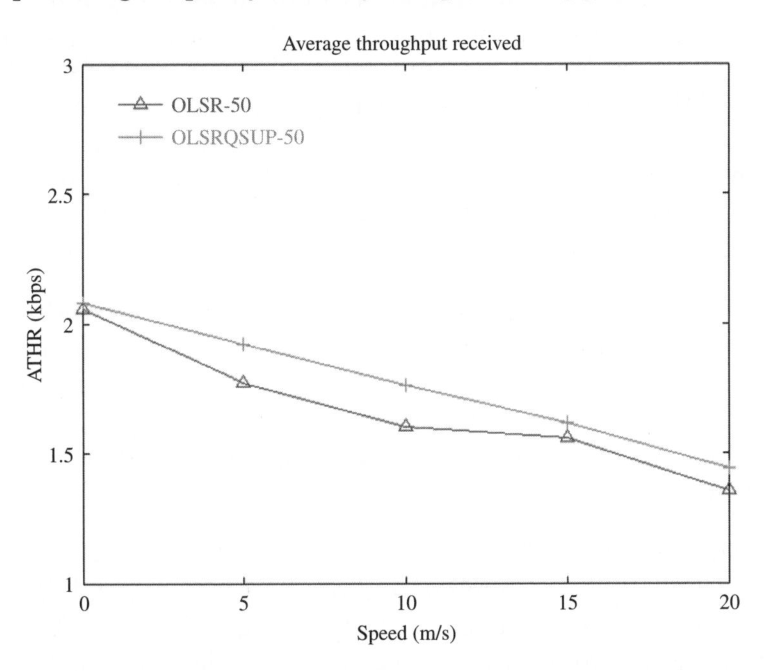

Figure 5.13. *ATHR for a 1,000 × 1,000 m network (50 nodes)*

Figure 5.13 shows the ATHR for a denser network. The throughput reaches 2.2 kbps in a static network, compared with 1.6 kbps for a less dense network of 30 nodes at the same velocity. In a denser network, more packets can be transmitted. In addition, from Figure 5.11, we notice that the delay offered by the OLSRQSUP is better. However, for the OLSR protocol, the reduced delay effect is offset by the higher number of packets transmitted (see PDR, Figure 5.7).

Moreover, we notice a degradation of this parameter as velocity increases. As nodes move, they will take longer to transmit traffic, and fewer packets will reach their destination.

5.3.3.2.4. Average number of MPRs

The average MPR number is defined as the number of MPRs generated in the network in the space of 5 s. Using the OLSR standard, TC messages are sent out every 5 s; only MPRs are responsible for the transmission of these messages. For this reason, we decided to look at the number of messages sent in this time interval and average out the value over the full duration of the simulation to calculate the average MPR number.

We note that the more nodes there are in the network, the more MPRs we find. In the case of a network with lower connectivity, a node may have less two-hop neighbors, so less need for MPRs. Thus, the denser the network, the higher the average MPR number, accompanied by an increase in control traffic.

The number of MPRs for a 30-node network using OLSRQSUP is very similar to that observed using OLSR for the same network density. The average MPR number in these cases is around 15 nodes, i.e. half of the network nodes, when the network is static. This can be explained by the fact that bandwidth in such situations (i.e. with zero mobility) is not particularly variable.

However, more than 50% of nodes in a 50-node network are used as MPRs. At velocity, $V = 0$ m/s, the OLSR protocol gives an average value of 38 MPRs, whereas the OLSRQSUP gives an average of 44. This difference is due to the modifications made to the MPR selection algorithm for the OLSRQSUP.

The OLSR protocol, in its original state, attempts to reduce control traffic and minimize MPR numbers to limit the number of TC messages. However, the OLSRQSUP protocol selects MPRs based on optimum bandwidth and delay parameters. This tends to produce a larger MPR group.

5.3.3.2.5. Control traffic

It seems evident that the denser the network, the more control traffic will be generated within the network. For a network using OLSR, the average control traffic is 12,000 and 16,000 packets for density of 30 and 50 nodes, respectively. As control traffic depends directly on network density, one might expect an increase of 60% between the two, i.e. around 20,000 packets for a 50-node network. The fact that this is not the case can be explained by the collision mechanism (Figure 5.9).

Moreover, we notice that there is less control traffic in a network with reduced mobility than in a network where nodes move at faster velocities. In networks where nodes move more slowly, less TC messages are required to maintain an up-to-date image of network topology. In a very mobile network, nodes change neighbors more frequently and so send more control messages to more neighbors, which in turn generate a large number of TC messages.

TC messages generated in networks of this kind are diffused by MPRs chosen using QoS constraints. It is therefore entirely natural for the OLSRQSUP protocol to generate more control traffic than its predecessor, OLSR.

It should be remembered that the OLSR protocol only generates TC messages to reflect network topology, whereas the OLSRQSUP needs to generate control messages to diffuse information on the QoS state of the network.

To avoid adding more control traffic to the traffic already generated by the protocol and extensions, each node checks for bandwidth changes every 0.5 s and updates its tables accordingly. It will only inform other nodes of this fact, however, if a time delay has expired. If a TC message was generated each time a change was detected in the available bandwidth or delay, a great deal of more traffic control would have been produced (although information on the bandwidth and delay parameters would certainly be more up-to-date).

If we reduce the number of HELLO and TC messages generated when change is detected, we run the risk of having imprecise network state information. We shall therefore keep control message generation at its current level, conforming to RFC 3626 parameters for emission intervals for these messages.

The OLSR protocol, in its original state, attempts to reduce control traffic and minimize MPR numbers to limit the number of TC messages. However, the OLSRQSUP protocol aims to select MPRs based on optimum bandwidth and delay parameters. This tends to produce a larger MPR group.

5.3.3.3. *Traffic impact*

So far, we have studied the performances of the two protocols by varying the network density and mobility parameters. Next, we shall consider another parameter, traffic, to observe the different behaviors of the protocols when faced with a variety of situations.

In evaluating the impact of traffic, we have chosen to use a static network model. Ad hoc networks are, of course, very mobile in nature, but in this case we wanted to observe the state of the network in terms of QoS if mobility is set at zero. The variable parameter in this case is the number of connections. Another change from the simulations carried out previously concerns network geometry. We will use a network with the same density in N/km^2, but with an architecture of 2,000 × 500 m so as to observe the impact of network topology on the two protocols.

5.3.3.3.1. Packet delivery ratio

The line graph in Figure 5.14 shows the variation in PDR depending on the number of connections established. We notice that the PDR declines as the number of connections increases, but the PDR value remains acceptable (over 80%). This value is higher than that given by the same network laid out as a grid rather than a strip, as fewer collisions occur in the latter layout due to the fact that nodes are less condensed.

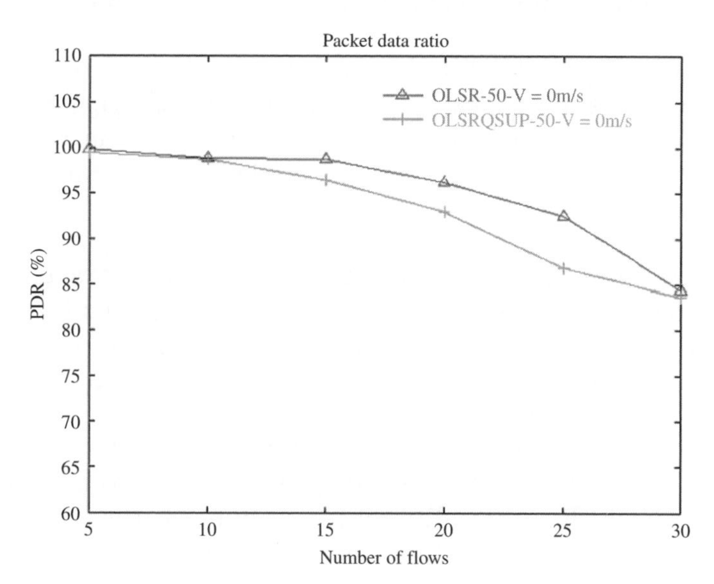

Figure 5.14. *PDR for a 2,000 × 500 m network*

We also note that the OLSR has a PDR around 5% higher than that produced by the OLSRQSUP. This difference is due to the fact that the OLSRQSUP generates more control traffic, which may compromise data transmission.

For this reason, the OLSR remains in the lead for the various fluxes. However, once the number of fluxes reaches 30 connections, the two lines converge. In any case, both protocols generally give an acceptable PDR.

5.3.3.3.2. Average packet delay

Figure 5.15 shows the APD depending on the number of connections. We observe a slight improvement in average delay (around 3 ms) for the OLSRQSUP protocol compared with the OLSR.

This improvement is weak if we wish to offer QoS guarantees to applications rather than simply best-effort routing. However, a number of factors help to explain the nature of this improvement.

First, the OLSRQSUP protocol produces more control traffic to ensure a good knowledge of network topology and QoS parameters. Packets therefore remain in the registers of intermediary nodes for longer before reaching their destinations.

If the number of fluxes is increased, more data packets will end up in the same buffer and will, consequently, have to wait there longer. For this reason, it is necessary to attribute different traffic classes to the fluxes in the network.

We also note that with 30 connections, average delay is 32 ms for the QoS version of the protocol, compared to 27 ms for the OLSR. This difference is due to the fact that OLSRQSUP and OLSR select paths in different ways.

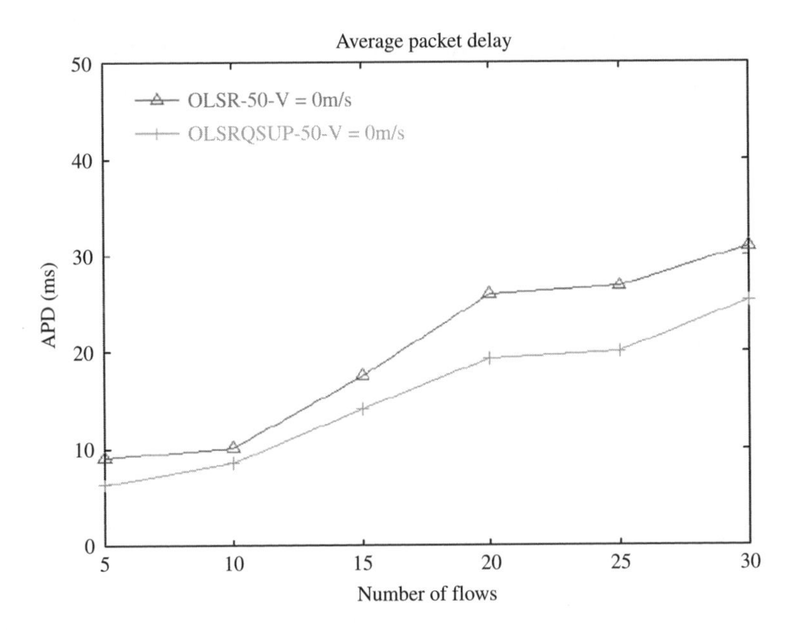

Figure 5.15. *APD for a 2,000 × 500 m network*

5.3.3.3.3. Average throughput

Figure 5.16 shows a variation in average throughput depending on the number of connections. The OLSRQSUP presents an improvement of around 10% on the values offered by the OLSR. If we look at the behavior of throughput in relation to the number of connections, we observe a slight improvement every five connections.

This is due to the fact that, with a linear network geometry, the node has fewer neighbors, so the total available bandwidth is shared among fewer nodes. However, once the number of fluxes reaches 20, throughput declines. Bandwidth is, in fact, shared between different network connections. We should remember that this parameter does not show the useful debit received by individual nodes, but the throughput from sending the first packet to reception of the last packet, averaged out over the number of connections. It is therefore clear that as the number of connections increases, the value of this parameter will decline.

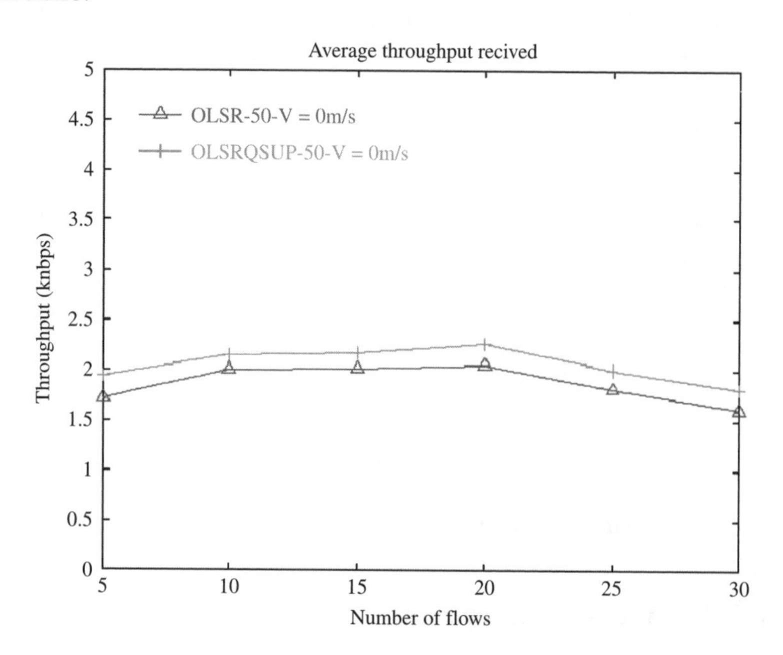

Figure 5.16. *ATHR for a 2,000 × 500 m network*

5.3.3.3.4. Average path length

This parameter is interesting as it allows us to evaluate the average number of hops a packet must make in this network topology to arrive at its destination. Generally, we might say that the shorter the paths, the better the performance. However, in making this statement, one forgets that the nodes through which a packet passes deal with variable levels of traffic and control messages.

Figure 5.17 shows that for all flux values, the OLSRQSUP protocol produces paths only one hop longer, on average, than those taken by the OLSR packets. This one-hop variation is due more to network topology than to the number of connections. To better understand performance in terms of average path length, more complex networks would need to be simulated, requiring considerably more processing power.

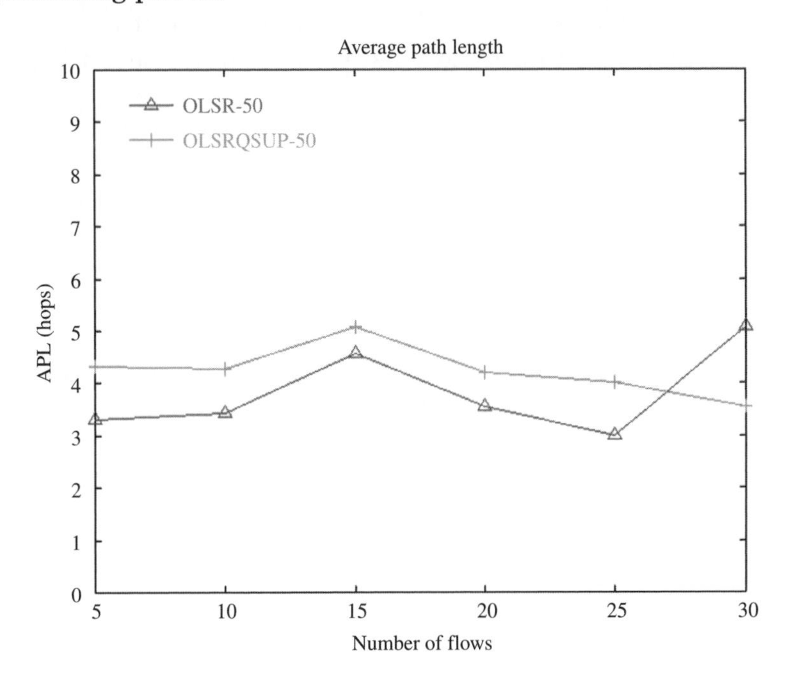

Figure 5.17. *Average path length for a 2,000 × 500 m network*

5.4. Conclusion

In this chapter, we have presented the implementation of the protocol and the results of simulations. We have noted that the traffic generated by the OLSRQSUP protocol is considerable and has an impact on performance. However, there are noticeable improvements in terms of delay and available bandwidth in paths used for data transmission.

The OLSR protocol enables the best-effort routing in a very dense network. It is able to support QoS extensions, allowing routing with QoS with a dispersion of fluxes across routes with maximum bandwidth and minimum delay. However, the TOH generated still requires improvement as it reduces the PDR. Consequently, the improvement of one communication parameter – mobility, traffic, topology, number of nodes, etc. – has an impact on the others, demonstrating the extreme sensitivity of ad hoc networks.

Chapter 6

Load Distribution in MANETs

6.1. Introduction

Mobile ad hoc networks (MANETs) are wireless networks that do not require a centralized administration or any fixed infrastructure. As mobile nodes communicate using radio channels of variable quality and reduced bandwidth, wireless links have lower capacity than wired links and, as a result, congestion is more of a problem. Furthermore, nodes in ad hoc networks are powered by batteries with a limited lifespan, reducing their storage capacity and processing power [BAS 04]. These constraints demand an even distribution of traffic across all nodes of a network. Otherwise, overloaded nodes may form a bottleneck, reducing network performance through congestion and long transmission delays, while rapidly consuming energy resources at node level, causing losses of connectivity in the network.

One solution to this problem is load balancing. Load balancing is defined as "the equitable distribution of processing and communication activities between different entities in a computer network to avoid overloading any one element" [ENC 07a]. This technique found favor following the exponential growth in Internet traffic over the last decade, mainly due to its capacity to cope with link failure and improve the quality of service (QoS) offered.

Unfortunately, load balancing is not present in shortest-path routing protocols for MANETs, and the bulk of traffic passes through the center of the network [PHA 02].

For this reason, we shall propose two load balancing mechanisms for this type of routing protocol. These work by pushing traffic away from the center of the network using the topological information available at node level. As network topology information at node level can be complete or partial depending on the routing protocol used (proactive or reactive), we distinguish between the two cases and propose solutions for both situations.

6.2. Previous approaches to the load-sharing problem

Although the mechanisms proposed above offer certain QoS guarantees, they do not guarantee an even distribution of load across the network. An admission control checks that sufficient bandwidth is available around each node used, guaranteeing sufficient bandwidth along the whole path taken by a packet.

Even if sufficient bandwidth is guaranteed along a route, this does not mean that intermediate nodes will not have to deal with high-traffic loads compared with other nodes in the network. It should be remembered that this effect is damaging to the network, even without congestion, as it rapidly consumes the resources (e.g. battery) of those "strategic" nodes most in demand to transmit traffic for other nodes. This unbalanced load distribution is a result of the shortest-path routing metric, which concentrates traffic around nodes located in the center of the network.

It has been demonstrated that, in MANETs, current routing protocols have a tendency to use certain nodes at the center of the network in a large number of paths [PHA 02]. Given that nodes have limited capacity, this tendency to impose excessive loads on certain nodes causes a number of problems.

First, the fact that several routes use certain nodes increases congestion in the nodes in question, reducing available bandwidth and increasing delays and packet loss rates following saturation of the nodes' buffers.

Moreover, through overuse, the nodes in question will rapidly run out of battery and will then no longer be connected to the network. As several routes depend on these nodes, a number of connections will be lost and connectivity "holes" will form. In the following section, we shall present certain load-balancing mechanisms suggested in existing publications on the subject.

In what follows, the word "load" will be used to designate the quantity of traffic dealt with by a node participating in one or more routes, defined as "the quantity of traffic received and transmitted by a node per unit of time on behalf of another node in the network."

Approaches to load balancing in existing literature can be grouped into two categories, depending on whether they use multipath or single-path routing. We shall present both approaches in sections 6.21 and 6.22.

6.2.1. *Approaches to load-sharing in multipath routing*

Using the multipath approach, more than one route is established between a source and a destination node pair, and traffic between these nodes is shared out across the different paths.

The concept of commutation or multipath routing has been discussed in different contexts. For example, circuit commutation in telephone networks uses a particular type of multipath routing, known as alternate-path routing. In this kind of routing, several paths – split into primaries and alternates – connect a source to a destination. Alternate-path routing was devised to reduce blockages and increase the global usage of the network. Effectively, as soon as a primary route (usually at one hop) between two communicators becomes unavailable due to a link failure or if maximum capacity has been exceeded, an alternate route – usually at two hops – is used to carry the surplus traffic (instead of blocking the traffic or directing it to another destination).

Multipath routing has also been used in data networks where connection-oriented transmission with QoS support is required. In the two previous examples, multipath routing has proved extremely useful in connection-oriented transmission networks where the probability of blockages is a major concern.

Multipath routing can also be used in packet transmission networks, such as the Internet, to reduce the load of the most in-demand nodes by redirecting traffic to nodes with lower demand.

This is the framework in which we find load-sharing for multipath networks. In addition to the tolerance for broken links, multipath routing allows traffic to be spread across several routes to reduce the load on individual links and eliminate the risk of bottlenecks.

To give an example, the asynchronous transfer mode transmission technique uses a signaling protocol, PNNI, to establish several routes between a source and a destination. As soon as the principal (or optimal) route becomes unavailable, alternate routes are used.

In this case, the approach to load balancing consists of finding the quantity of traffic to transmit through a path that minimizes a certain cost function. In what follows, we present an overview of the work carried out by Yin and Lin [YIN 04] of the University of Tsinghua, Beijing, in which the researchers tried to provide optimal load distribution for MANETs in the context of multipath routing.

6.2.1.1. *The multipath adaptative load balancing (MALB) mechanism*

Yin and Lin propose a load-sharing mechanism known as MALB [YIN 04].

In truth, their contribution is not a routing protocol, but a mechanism able to integrate any multipath routing protocol.

To do this, the authors presume the existence of several separate paths between a pair of nodes. The MALB mechanism operates at source node level and aims to distribute traffic across the different available routes based on a statistical study.

6.2.1.1.1. Modeling of the network

The network is represented by a graph of which the apices N are the mobile nodes in the network. W is the group of node pairs (source and destination) indexed from 1 to w. For each pair $w = (i; j)$, P_w is the group of separate routes leading from i to j, and r_w the total flow of traffic transmitted from i to j and divided into several flows x_{wp} for each path p of P_w, so that:

$$\sum_{p \in P_w} x_{wp} = r_{wp} \qquad [6.1]$$

From another angle, let x_n be the flow arriving at any node in the network, n. The identity of conservation is as follows:

$$x^n = \sum_{w \in W} \sum_{p \in P_w, n \in p} x_{wp} \qquad [6.2]$$

Finally, the authors attribute a cost function $D_n(x^n)$ to each node for the traffic x^n. Looking at the end-to-end delay, they take delay as the cost function, considering it as an M/M/1 queue:

$$D_n(x^n) = \frac{1}{(\mu_n - x^n)} \qquad [6.3]$$

where μ_n is the capacity of the node n.

6.2.1.1.2. Formulation of the problem

With the objective of minimizing the cost $D(x) = \sum_{n \in p} D_n(x^n)$ sharing traffic in an optimal manner between the paths p of P_w, the problem can be written as follows:

$$\text{minimize } D(x) = \sum_{n \in p} D_n(x^n) \qquad [6.4]$$

with the constraints:

$$\sum_{p \in P_w} x_{wp} = r_{wp}, \forall w, p \qquad [6.5]$$

$$x_{wp} \geq 0, \forall w, p \qquad [6.6]$$

Given that the vector x_w is only optimal if, for each path p of P_w, the values $(\partial D(x)) / (\partial x_{wp})$ are equal, Lin *et al.* use this value as a measure of congestion in the MALB mechanism. As for the aforementioned problem of minimization under constraints, they use the gradient algorithm to determine the step γ, defined as:

$$x_{wp}(t+1) = x_{wp}(t) - \gamma \frac{\partial D(x)}{\partial x_{wp}} \quad [6.7]$$

This guarantees the convergence of the sequence $x_{wp}(t+1) - x_{wp}(t)$ when t tends to $+\infty$.

6.2.1.1.3. Description of the mechanism

The MALB mechanism obliges each node to save traffic arrivals periodically. For this, the source sends out probe packets periodically over each separate path. On receiving a probe packet, a node sends an evaluation of the traffic traversing it $x_{wp}(t)$ back to the source.

At the moment of transmission, the source estimates the traffic running over each link contingent on previously received traffic reports using equation [6.6].

6.2.1.1.4. Performance evaluation

Load-sharing in multipath routing has the advantage of coping well with link rupture (fault tolerance) and of reducing the load of different routes linking a node pair.

Nevertheless, the mechanism also has its disadvantages. The use of several routes for the same destination occupies more space in the routing table, using up node memory capacity and increasing the size of packet headers, a major issue in a MANET.

In the MALB mechanism presented above, we have seen that each source receives periodic reports from all other sources on the state of traffic, reports which are saved to estimate the best route to take for future transmissions.

However, the choice of paths using the end-to-end delay in the network also resolves the problem of minimization under constraints. But although the proposed method (gradient method) is fairly simple (offering resolution in linear time), the time taken remains considerable in the case of dense networks. The MALB protocol thus presents scalability problems.

Finally, the MALB mechanism, like most load-sharing mechanisms for multipath routing, is based on the hypothesis that enough separate routes exist between a pair of nodes. Routes of this kind are rare in MANETs, and their detection is even more difficult.

6.2.2. *Approaches to load-sharing in single-path routing*

In the single-path approach, as the name suggests, only one route is established between a source/destination pair of nodes.

Several mechanisms of this kind have been suggested, including the use of different routing metrics [WAN 00], packet caching [VAL 03], directional antennae [ROY 02], etc.

To give an example, [ZHO 01] presents a new routing protocol, LBAR, where the routing metric takes the degree of nodal activity into account, defined as the number of active paths established through the node.

Elsewhere, [LEE 05] suggests that overloaded nodes should have the possibility of refusing to participate in new paths for as long as their load is not reduced. To this end, each node would define an overloading threshold, allowing it to decide whether to accept or reject each new request to participate in a connection.

6.2.3. *Performance comparison of load-sharing approaches in single-path and multipath routing*

We have given above the two approaches to load-sharing to be found in existing publications. The question of whether one is more effective than the other now needs to be addressed. The issue has already been studied in the case of wired networks.

Research on this subject in the radio domain has only just begun. Only one document published by Pham and Perreau [PHA 02] of the University of South Australia gives an analytical model for comparing the performances of these two types of routing in MANETs.

In the following section, we compare the two approaches from an optimal load-sharing standpoint, using a summary of the results of this research.

6.2.3.1. *Packet queues in single-path routing*

In this section we compare the number of packets waiting in network node buffers in a single-path routing situation with the same value for multipath routing. We then identify conditions where multipath routing offers shorter queues, and thus a better distribution of load across the network.

It has been proven [PHA 02] that the total traffic traveling across a node located at distance r from the center of the disc is given by equation [6.8]:

$$\lambda(r) = (\pi R^2 \delta - 1)\lambda + \frac{\pi(R^2 - r^2)^2 \delta^2 \beta}{2}\lambda \qquad [6.8]$$

where δ is the area density of the nodes, λ the level of internodal transmission, R the radius of the disc around the group of nodes, and β a constant allowing for the fact that the paths between two nodes are not completely straight.

Details of the proof of this result can be found in section 6.3.

Using Little's theorem, we can deduce the number of packets in an M/M/1 queue for a node at distance r from the center of the disc:

$$N_{\text{pac}}(r) = \frac{\lambda(r)}{\eta - \lambda(r)} \qquad [6.9]$$

where η is the processing capacity of the queue.

From the previous equation, we can work out the total number of packets waiting in the network:

$$N_{\text{pac,total}} = \int_0^{2\pi} 2\pi r N_{\text{pac}}(r)\,\mathrm{d}r \qquad [6.10]$$

The average number of packets in a queue is given by:

$$N_{\text{pac,unique}} = \frac{1}{\pi R^2 \delta} \int_0^{2\pi} 2\pi r N_{\text{pac}}(r)\,\mathrm{d}r \qquad [6.11]$$

6.2.3.2. *Packet queues in multipath routing*

A load-sharing algorithm in a multipath routing protocol distributes traffic in an equitable manner among all the nodes in the network. For this reason, packets experience a lower end-to-end transmission delay.

Let L_m, λ_m, and η be the average length of a route, the level of internodal traffic, and the processing capacity of a node, respectively. As the number of nodes in the network is $\pi R^2 \delta$, it is clear that the number of possible connections in the network must be $(\pi R^2 \delta - 1)\ \pi R^2 \delta$.

Consequently, the average traffic in the network is $(\pi R^2 \delta - 1)\ \pi R^2 \delta L_m \lambda_m$ and thus $(\pi R^2 \delta - 1)\ L_m \lambda_m$ per node. Following Little's theorem, the average number of packets in the queue of a network node using multipath routing is then given as follows:

$$N_{\text{pac,multiple}} = \frac{(\pi R^2 \delta\text{-}1) L_m \lambda_m}{\eta - (\pi R^2 \delta\text{-}1) L_m \lambda_m} \qquad [6.12]$$

6.2.3.3. *Commentary*

According to the previous study, to guarantee that load-sharing in multipath routing offers a better distribution of traffic than single-path routing and thus lower levels of congestion, the former must give a lower number of packets per queue than the latter, i.e. $N_{\text{pac, multiple}}$ must be lower than $N_{\text{pac, unique}}$.

To this end, the average length L_m of a path in multipath routing must satisfy the following condition:

$$L_m < \frac{\eta N_{\text{pac,unique}}}{(N_{\text{pac,unique}} + 1) - (\pi R^2 \delta - 1)\lambda_m} = L_{\max} \qquad [6.13]$$

From this, if $L_m > L_{\max}$, the use of a load-sharing mechanism in a multipath routing protocol is no more effective than using the same mechanism in a simple single-path routing protocol.

Furthermore, [PER 00] demonstrates that load balancing by multipath routing is no more effective than approaches used in single-path routing unless the multiple paths all use separate nodes, something which is difficult to find in a MANET [PER 00]. If the multiple paths are not completely separated, [GAN 04] shows that a large number of paths must be available for a noticeable improvement in load distribution, typically around 100 different paths between a node pair in a network of 500 nodes.

Despite the considerable number of load balancing mechanisms proposed, we still do not possess an efficient mechanism for the equitable distribution of traffic between nodes in a MANET. On the one hand, load balancing solutions using multipath routing are not suitable for use in a MANET as they require the periodic exchange of a large number of reports between nodes concerning their respective loads. On the other hand, the approaches proposed using single-path routing, while effective in reducing congestion in overloaded nodes, do not guarantee an even distribution of load over different network nodes.

To provide an effective load-sharing mechanism, we shall first study load distribution in a MANET to find out why certain nodes are overloaded and not others. We shall then look for a way to reduce the traffic load of these nodes and increase that of underused nodes, while taking account of the constraints of a MANET (e.g. minimum exchange of topology data messages).

6.3. Analytical study of the load-sharing problem in an ad hoc network with shortest-path routing

To study the load-balancing problem in an ad hoc network, we shall reuse the model presented in [PHA 02] to analyze the distribution of traffic in a network managed by a routing protocol with a shortest-path routing metric.

Let us consider a network containing N nodes spread across a disc D of radius R and center O. We shall suppose that mobile nodes are distributed uniformly across the surface of D with an area density δ. N is then linked to area density by

$$N = \pi R^2 \delta \qquad [6.14]$$

Each link between two nodes is described by a transmission debit expressed in packets per second.

Let A be a node located at distance r from the center of the disc and $X(\alpha)$ a point on the circumference of the disc so that the angle formed by the rays [AO) at [$AX(\alpha)$) is α. Let $S_\alpha(\mathrm{d}\alpha)$ be a portion of the disc D with vertex A and angle $\mathrm{d}\alpha$. This network model is shown in Figure 6.1.

The aim is to calculate the number of routes traveling through A and of which the sources are points of $S_\alpha(\mathrm{d}\alpha)$. To respond to the question, we need to know the "destination surface," i.e. the portion Δ of disc D containing all of the points B which may be the destination of traffic generated by the points of $S_\alpha(\mathrm{d}\alpha)$ and passing through A.

As the routing protocols studied using the shortest-path metric, we can assimilate optimal routes to straight lines following the hypotheses of the study (i.e. that network nodes are presumed to be distributed with uniform density over the whole disc).

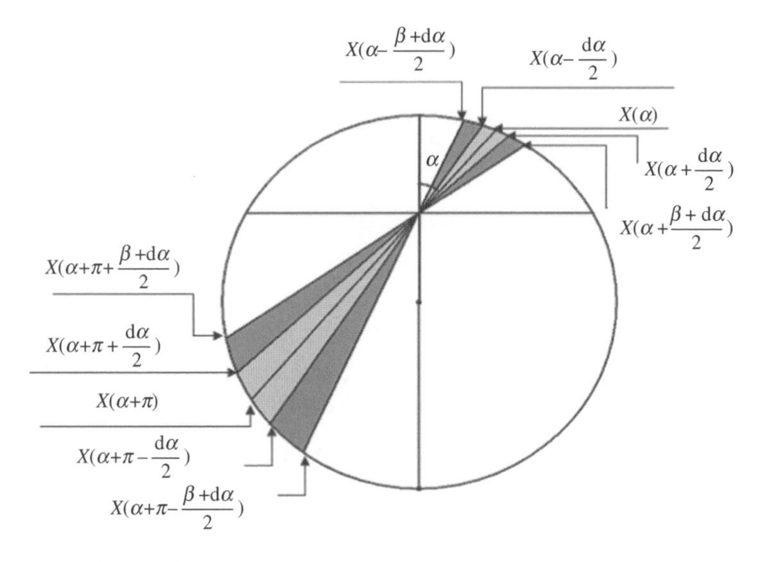

Figure 6.1. *Analytical model of Pham and Perreau [PHA 02]*

If optimal routes were perfectly straight lines, then Δ would be the portion of the disc $S_{\alpha+d_\alpha}(\mathrm{d}_\alpha)$ and opposed by the vertex A to $S_\alpha(\mathrm{d}\alpha)$. In reality, routes are not perfectly aligned as distribution is discrete and random, so the surface of Δ will be larger, so:

$$\Delta = S_{\alpha+\pi}(\mathrm{d}\,\alpha + \beta) \tag{6.15}$$

where β is a small positive real number independent of α. The value of β depends on the density of the network and the distribution of nodes. Its value may be determined by a graphical study of node distribution or by simulation [PHA 02].

We shall now calculate the values of $S_{\alpha+d\alpha}$ (d_α)and $S_\alpha(\mathrm{d}\alpha)$. As $\mathrm{d}\alpha$ is infinitesimal, we can use the following approximations:

$$\sin(\mathrm{d}\,\alpha) \approx \mathrm{d}\,\alpha \tag{6.16}$$

$$AX(\alpha - \mathrm{d}\,\alpha) \approx AX(\alpha + \mathrm{d}\,\alpha) \approx AX(\alpha) \tag{6.17}$$

$$S_\alpha(\mathrm{d}\,\alpha) = \frac{AX(\alpha - \mathrm{d}\,\alpha) \cdot AX(\alpha + \mathrm{d}\,\alpha) \cdot \sin(\mathrm{d}\,\alpha)}{2} \tag{6.18}$$

This produces the following results:

$$S_\alpha(\mathrm{d}\,\alpha) = \frac{AX^2(\alpha)\,\mathrm{d}\,\alpha}{2} \tag{6.19}$$

$$S_{\alpha+\pi}(\mathrm{d}\,\alpha + \beta) = \frac{AX^2(\alpha + \pi)d(\alpha + \beta)}{2} \tag{6.20}$$

We have also presumed that nodes are evenly distributed across the disc. Consequently, the number of nodes in $S_\alpha(\mathrm{d}\alpha)$ and $S_{\alpha+\pi}(\mathrm{d}\alpha + \beta)$ will be $\delta \cdot S_\alpha(\mathrm{d}\alpha)$ and $\delta \cdot S_{\alpha+\pi}$ $(\mathrm{d}\alpha + \beta)$, respectively.

An optimal route (following a shortest-path routing algorithm) which traverses A being a route which links a point of $S_\alpha(d\alpha)$ to a point of $S_{\alpha+\pi}$ $(d\alpha + \beta)$, the number of optimal routes traveling through A is as follows:

$$N = S_\alpha(\mathrm{d}\,\alpha) \cdot S_{\alpha+\pi}(\mathrm{d}\,\alpha + \beta) \cdot \delta^2 \qquad [6.21]$$

$$N = \frac{AX^2(\alpha+\pi) \cdot AX^2(\alpha) \cdot \delta^2 \cdot \mathrm{d}(\alpha)\mathrm{d}(\alpha+\beta)}{4} \qquad [6.22]$$

Leaving out α in front of β, we obtain:

$$N = \frac{AX^2(\alpha+\pi) \cdot AX^2(\alpha) \cdot \delta^2 \cdot \beta \cdot \mathrm{d}(\alpha)}{4} \qquad [6.23]$$

To calculate the lengths $AX(\alpha)$ and $AX(\alpha + \pi)$, we use the fact that the triangles ABB_1 and ACC_1 illustrated in Figure 6.2 are similar.

So:

$$AC.AB = AC_1.AB_1 = R^2 - r^2 \qquad [6.24]$$

Replacing $AX(\alpha)$ and $AX(\alpha + \pi)$ by AB_1 and AC_1, respecttively, we obtain:

$$N = \frac{(R^2 - r^2) \cdot \delta^2 \cdot \beta \cdot d(a)}{4} \qquad [6.25]$$

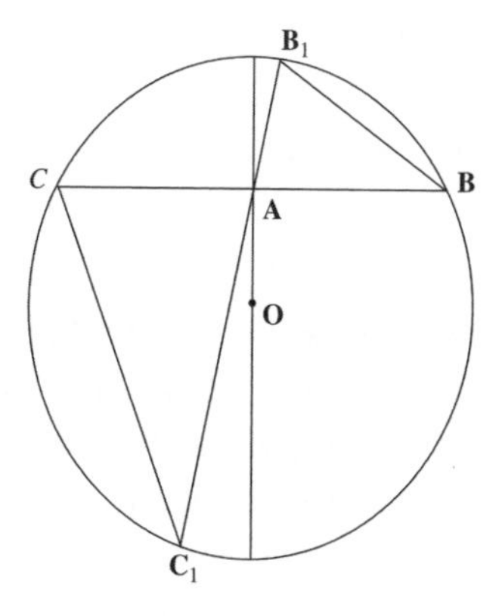

Figure 6.2. *Similarity of triangles ABB1 and ACC1*

Taking bidirectional links into consideration, the total number of optimal routes in the network passing through A is:

$$N_A = 2 \times \int_0^{\pi} \frac{(R^2 - r^2) \cdot \delta^2 \cdot \beta}{4} \mathrm{d}(\alpha) \qquad [6.26]$$

$$N_A = \frac{\pi(R^2 - r^2) \cdot \delta^2 \cdot \beta}{2} \qquad [6.27]$$

and the traffic relayed by A will be:

$$\lambda_A = \lambda \cdot N_A \qquad [6.28]$$

$$\lambda_A = \lambda \cdot \frac{\pi(R^2 - r^2) \cdot \delta^2 \cdot \beta}{2} \qquad [6.29]$$

However, a node A does not just transmit traffic for other nodes. It also generates its own traffic, which it transmits to other nodes in the network. As the disc contains $\pi R^2 \delta$ nodes, node A can communicate with a maximum of $(\pi R^2 \delta - 1)$ nodes and the total traffic transmitted by a node at distance r from the center of the disc becomes:

$$\lambda(r) = \lambda \cdot \frac{\pi(R^2 - r^2) \cdot \delta^2 \cdot \beta}{2} + \lambda \cdot (\pi R^2 \delta^2 - 1) \qquad [6.30]$$

We note that this traffic is inversely proportional to the distance r from the center of the disc. Traffic load is not uniform in an ad hoc network using a shortest-path routing algorithm. Maximum load is found at the center of the network and decreases the further we move from the center.

From this study, we can conclude that traffic load is highest at the center of a MANET using a shortest-path routing algorithm. Consequently, the way to guarantee load balancing would be to find a mechanism to push load away from central nodes. We shall attempt, in the following section, to propose two load-sharing mechanisms based on this idea.

6.4. Proposition

According to the previous study, the shortest-path metric is at the root of the load distribution problem in ad hoc networks. By favoring the shortest routes, the shortest-path routing protocol overloads nodes at the center of the network, creating a bottleneck that limits network capacity.

For this reason, a new metric is needed to push traffic away from the center of the network. This metric must, first and foremost, be able to identify the central nodes.

Due to the diversity of shortest-path routing protocols in MANETs, we shall propose two different approaches to load balancing: one for proactive protocols (where the metric seeks the center of a network of which the topology is known) and one for reactive protocols (where the metric seeks the center of a network of unknown topology).

6.4.1. *Proactive routing protocols*

This is easier of the two cases, as *a priori* the topology of the network is known at any given time by all nodes in the network. We have chosen to study the optimized link-state routing (OLSR) protocol [CLA 03], but our findings can be extended to any other proactive routing protocol using the shortest-path routing metric.

We shall model a MANET formed of N mobile nodes by a connate[1] nonoriented[2] graph $\mathcal{G} = (\mathcal{V},\zeta)$, where $\mathcal{V}$ denotes the group of vertices representing the network nodes and ζ is the group of lines linking these vertices which represent the links between the nodes. We shall use the following definitions and theorems, taken from graph theory [EME 07, WEI 07]:

DEFINITION 6.1.– Let the chain $\mathcal{G} = (\mathcal{V},\zeta)$ be a finite sequence of vertices, so that:

$$\forall\ 0 \leq j < m,\ (s_{i+j}, s_{i+j+1}) \in \zeta$$

The integer m is the length of the chain, written $m= l(c)$.

DEFINITION 6.2.– If s and s' are two vertices on the graph, then:

$$d(s,s') = \begin{cases} \infty, & \text{if} \quad \{c, c'\text{chain from } s, s'\} = \varnothing \\ \min\{l(c)c' \text{ chain from } s, s'\}, & \text{otherwise} \end{cases}$$

THEOREM 6.1.– If $\mathcal{G} = (\mathcal{V},\zeta)$is connate then the function d: $\mathcal{V}^2 \rightarrow \mathbb{R}$ is a distance on $\mathbb{R}$, where $\mathbb{R}$ designates the group of real numbers.

DEFINITION 6.3.– A geodesic from s to s' is a chain from s to s' of length d(s,s').

---

1 The graph is connate as there is a path between each pair of nodes. In this case, each node within the network is connected to at least one other node.

2 The graph is nonoriented as, in OLSR, links are bidirectional and symmetric.

DEFINITION 6.4.– The diameter $\mathcal{D}(\mathcal{G})$ of a connate graph $\mathcal{G}$ is the length of the longest geodesic of $\mathcal{G}$.

DEFINITION 6.5.– The eccentricity $\mathcal{E}(s)$ of a vertex s is:

$$\mathcal{E}(s) = \max \{\mathrm{d}(s,s'),s' \in \mathcal{V}\}$$

From the definitions given above, we obtain a mathematical characterization of the center of a connate nonoriented graph.

DEFINITION 6.6.– A center of $\mathcal{G}$ is a vertex s at minimum distance from the center, i.e. a vertex s where:

$$\mathcal{E}(s) = \min \{\mathcal{E}(s'), s' \in \mathcal{V}\}$$

This same definition characterizes the center of a MANET as described in the analytical study above. The distances defined in routing metrics are given as a number of hops, corresponding perfectly with the definition of the distance between two vertices on the graph (Definition 6.2). Moreover, the definition of central nodes using an eccentricity value corresponds to the definition of the center of a network as presented in the analytical study, as nodes closer to the center are also closer to other nodes and so, using a shortest-path metric, most established paths will pass through them.

Note that this characterization is very practical for the OLSR. As the protocol is proactive, each node has access to information on what paths to take to any destination within the network at any given moment. Within this information we find the R_dest field, which gives an estimation of the number of hops separating the source from the destination [CLA 03]. Subsequently, any node in a MANET using the OLSR routing protocol can calculate its eccentricity fairly quickly[3]: this distance is *the largest number of hops separating the node from a destination in the network.*

Subsequently, the node must simply inform other nodes of its distance from the center so they can determine in what measure the node belongs to the central group, by comparing this value with those of other nodes. This operation is described in detail below.

6.4.1.1. *Proposal for a new routing metric*

Using the definitions and theorems given in the previous section, we propose that each node should aim to reach destinations at more than two hops using paths that are not *a priori* the shortest, but with the least load.

We judge that immediate neighbors and two-hop neighbors are too close to the source for a better route than the shortest path available. We suggest that the optimality of these routes in comparison with the shortest path should be assessed using the equation [6.31]:

$$\min \frac{1}{n} \sum_{i=1}^{n} \varepsilon(i) \qquad [6.31]$$

where n is the number of hops on the path and i a node on the path with eccentricity $\varepsilon(i)$. This minimization is simple and free of constraints. First, the number of hops separating the two specified nodes cannot be zero. Second, the retroactive action of each term in the cost function on the other means that the minimum always exists: a shortest-path route not only minimizes n (so maximizes the term) but

3 The complexity of the calculation is $O(n)$, where n is the size of the routing table. All that needs to be done is to find the entry in the table that has the highest R_dest value.

also minimizes the sum of the distances of each node from the center, as these nodes belong to the central group.

6.4.1.2. *Modification of the routing algorithm*

In accordance with RFC 3626 [CLA 03], upon receiving a topological data update, a node using the OLSR protocol compares the routes announced by its neighbors with those present in its routing table and implements the necessary updates. Using our load-sharing mechanism, the comparison of old and newly announced routes should be carried out using the metric given in equation [6.31].

This metric brings into play the distance of a node from the center of the network, a value defined as the largest distance (in terms of number of hops) separating the node from other nodes in the network.

This metric is easy to calculate in OLSR. RFC 3626 requires that nodes begin by saving data on one-hop neighbors in their table, followed by two-hop neighbors, three-hop neighbors, etc. Consequently, the distance between the node and the last input in its routing table gives the distance of the node from the center, as defined in graph theory.

Once the distance of the node from the center is known, the local node should share this information with other network nodes. This information will be included in the HELLO message in the same way as QoS extensions proposed by the QOLSR protocol [BAD 00].

Furthermore, as RFC 3626 limits the maximum distance to 255 hops (maximum value of the Time To Live field in an OLSR packet), this information will be contained in one byte and will therefore not have a noticeable effect on levels of signalization traffic in the network.

Upon receiving a HELLO message sent from node *B,* node *A* takes the distance-from-center value of node *B* given in the HELLO message. *A* then recalculates the average distance from center of all the paths in which node *B* participates and which are present in its topological data table. *A* then compares these routes to those present in its routing table.

If a path to a destination has a lower distance-from-center value than that of the path to the same destination contained in the routing table, then *A* will replace its existing table entry with the new path.

Algorithm 6.1 illustrates this routine.

ALGORITHM 6.1.– Proposed routine for identifying less-loaded routes.

For any route *route_k* , k^{e} entry in the *topology database*
  If *B* appears in *route_k* **then**
    For any route *route_j*, j^{e} entry *routing table*
      If (same_destination(*route_k, route_j*) and
        eccentricity(*route_k*)< eccentricity (*route_j*)) **then**
          Replace *route_j* by *route_k*
      End If
    End For
  End If
End For

6.4.2. *Reactive routing protocols*

In ad hoc on-demand distance-vector (AODV), nodes only have routes to a section of the other nodes in the network. It is therefore no longer possible to use the same notion of eccentricity used in the previous section, as the mathematiccal definition is based on a calculation of the distance separating each node from all the others in the network. A new characterization of the center of the network is needed, which must conform to the objectives of RFC 3561. Specifically, this characterization must not require the diffusion of new control messages within the network or necessitate localization.

In studying the route discovery mechanism used in the AODV, we observe that, when a route request (RREQ) message is received by an intermediary node, this node first checks its routing table to see if the information contained in the message is more recent than that contained in the table, and, if this is the case, updates the routing table with the new data. It follows that nodes in the center of the network will have larger routing tables than peripheral nodes as they receive a greater number of RREQ messages. This would suggest that the size of the routing table could be used to measure the degree of proximity of a node to the center of the network.

Furthermore, in AODV, information stored in routing tables has a limited lifespan. Entries in the routing table expire once their predetermined lifespan is exceeded. This mechanism was introduced to guarantee the freshness of routing information at node level [PER 03]. Consequently, the size of the routing table would not only give the distance of a node from the center of the network, but also guarantee the recency of this measurement.

This intuitive reasoning fits in with the results of our previous analytical study. We demonstrated that the number of possible shortest-path routes passing through a node A is:

$$N_A = \frac{\pi(R^2 - r^2) \cdot \delta^2 \cdot \beta}{2} \qquad [6.32]$$

This gives the number of destinations which can be reached by node A and is also, by definition, the size of node A's routing table.

We shall attempt to find the relationship between this value and the end-to-end delay in the following. From this, we will be able to produce a routing metric that guarantees load balancing.

According to [YIN 04], the queue of a node in an ad hoc network behaves in the same way as an M/M/1 queue. Consequently, the approximate delay experienced by traffic at node A can be given by equation [6.33] [YIN 04]:

$$D(r) = \frac{1}{\mu - \lambda(r)} \quad [6.33]$$

where μ is the capacity of the link over which traffic is received by A. The traffic relayed by A is given by equation [6.34]:

$$\lambda(r) = N_A \cdot \lambda \quad [6.34]$$

Thus, we obtain the required relationship between the size of the routing table and the delay:

$$D(r) = \frac{1}{\mu - N_A \cdot \lambda} \quad [6.35]$$

We have therefore determined the relationship between the size of the routing table of a node and the delay at the level of the same node by analytical means: the smaller the routing table (i.e. the further the node is from the center of the disc), the better (i.e. lower) the end-to-end delay. We may now define the load-balancing mechanism properly using the results of these calculations.

6.4.2.1. *Proposed new routing metric*

To be able to choose the least burdened route, a destination node D must no longer ignore late RREQ messages, as different RREQ messages do not necessarily arrive simultaneously (each having, *a priori*, taken a different route to the others). This being the case, node D needs criteria that will allow it to choose the least loaded of all the proposed routes. Following on from our work in the previous section, we propose equation [6.36]:

$$\min \sum_{i=1}^{n} \text{routing table size} \qquad [6.36]$$

where *routing_table_size(i)* is the size of the routing table of a node participating in the route under consideration. The division by n, number of hops, means that the metric takes the number of hops on the path into account to correctly evaluate the load carried over a route. In addition, this choice allows us to minimize the size of data to include in the RREQ message, which itself impacts upon load.

6.4.2.2. *Proposed handover mechanism*

We have defined a routing metric that allows a destination node to choose a route from those received through RREQ messages. However, these RREQ messages do not arrive simultaneously, as *a priori* they follow different paths.

Let us suppose that D receives a late RREQ message announcing a better route at a point when data are already being received from S. A mechanism is required to allow D to establish the new route to S and for S to move the traffic onto this new path. This situation is similar to the one observed in cellular networks, traffic handover [YIN 04]. The traffic handover mechanism allows a mobile node to change base station in the course of communications to balance traffic load between cells [YIN 04].

For these reasons, we propose a mechanism for traffic handover between nodes in a wireless network. On receiving a late RREQ message containing a less-loaded route, the destination node D will set up a second route to S following the reverse path setup process proposed in RFC 3561 [PER 03].

Once the second route is established, S ceases transmission over the original path and instead sends packets via the new path.

6.4.2.3. *Modification of the routing algorithm*

We need to redefine the method of transportation of RREQ messages and the way in which route reply messages are generated. Note that in AODV, routing tables are updated when RREQ messages are received. Before replying to an RREQ message, the receiving node consults the contents of the message and finds the list of its precursors along the path.

If this list provides a shorter route to the source of the message than that in the node's routing table, the node updates the routing table by replacing the old route to the source of the RREQ message with the new route received.

The average eccentricity of a route is calculated as follows:

– The source S sends an RREQ message including the size of its routing table as the eccentricity value of the route:

$$\varepsilon_0 = \text{routing\_table\_size}\ (S)$$

– The neighboring node V_1, not possessing a route to the destination, diffuses the RREQ message with a new eccentricity value:

$$\varepsilon_0 = \frac{1}{2}\left(\text{routing\_table\_size}\ (S) + \text{routing\_table\_size}\ (V_1)\right)$$

– The process is repeated. An intermediate node V_{n-1}, not possessing a route to the destination, diffuses the RREQ message with a new eccentricity value:

$$\varepsilon_0 \frac{1}{n}\left(\text{routing\_table\_size } (S) + \sum_{i=1}^{n-1} \text{routing\_table\_size } (V_i) \right)$$

From this point on, the proposed route selection algorithm is similar to that proposed for the OLSR. The new route for comparison with existing routes is not, however, contained in the topology database (no structure of this kind exists for the AODV), but in the RREQ messages themselves. Algorithm 6.2 gives the proposed method for routing table updates in the AODV.

ALGORITHM 6.2.– *Route comparison by the new routing metric*

If (*RREQ* received) then
  Read announced route[4] *route_RREQ* in RREQ
  Read average eccentricity of *route_RREQ*
  For any route *route_j,* j^e entry in the *routing table*
    If (same_destination(*route_k,route_RREQ*) and
      eccentricity(*RREQ*)<eccentricity(*route_j*)) **then**
        Replace *route_j* with *route_k*
    End If
  End For
End If
End For

6.5. Performance evaluation of proposed load-balancing mechanisms

We added our proposed load-balancing mechanisms to implementations of the AODV and the OLSR under ns-2.29 and ns-2.27, respectively [NS 07], to visualize the performance of our optimizations.

---

4 The precursor list, see RFC 3561.

We subsequently envisaged 10 different scenarios for simulation, each distinguished by different traffic and mobility parameters. The mobility model used was random waypoint [NS 07].

The pause time in our simulations varied from 0 to 700 s and node velocity varied from 0 to 20 m/s with an average of 10 m/s to study the performance of our load-balancing mechanisms at both high and low mobility.

The simulated network was made up of 50 nodes of which 40 are traffic sources. The simulated traffic had a constant bit rate (CBR) of 2 Mbits/s and packet size was 512 bytes. The surface occupied by the nodes was 670 × 670 m and the simulations lasted 900 s. This duration was chosen as being sufficiently long to guarantee the stability of results and a fairly narrow confidence interval (10%, i.e. a confidence level of 90% for each point on the graphs presented).

The multiple access collision layer protocol was IEE 802.11b with a nominal debit of 11 Mbit/s. To better show the use of the proposed mechanisms, we increased the average number of hops on a route by limiting the range of nodes to 50 m.

The performance of the routing protocol with the added load-balancing mechanism is expressed using the following terms:

– average load distribution in relation to distance from the center of the network;

– average end-to-end delay; and

– packet delivery ratio (PDR) (or packet delivery fraction).

6.5.1. *Evaluation: load distribution*

Figures 6.3 and 6.4 show load distribution in the network in relation to distance from the center of the network for both the OLSR and the AODV. Load was defined as "the quantity of traffic received and transmitted by a node per unit of time" and expressed in Mbit/s. We notice a high concentration of data traffic in the center of the network even though the traffic models used were uniform (i.e. nodes all produce packets at the same rate).

We have therefore demonstrated the unbalanced distribution of load in a MANET using OLSR or AODV.

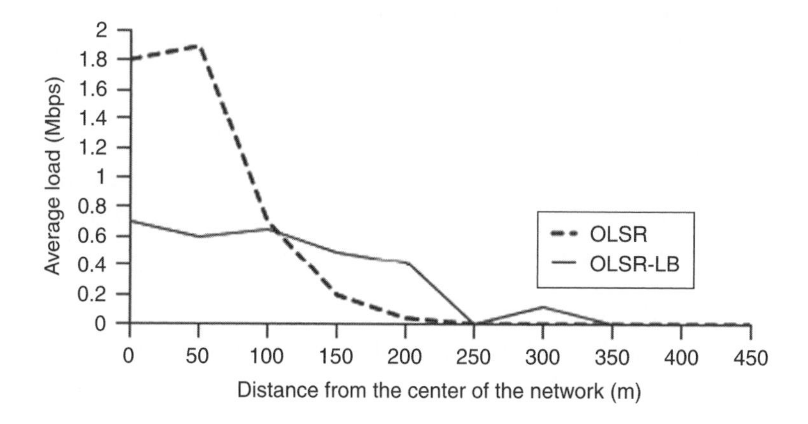

Figure 6.3. *Load distribution and eccentricity in OLSR*

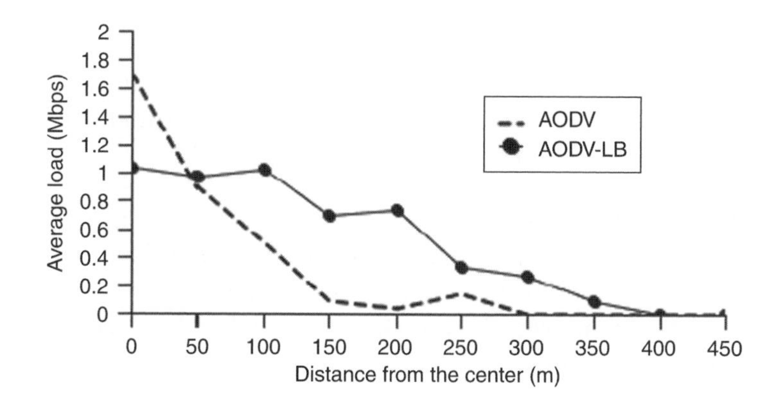

Figure 6.4. *Load distribution and eccentricity in AODV*

The analytical study of load distribution presented in the previous section linked the concentration of traffic at the center of the network to the use of shortest-path routing algorithms, as routes created by this metric pass through central nodes. The use of modified routing metrics alleviates the load in the center of the network and increases the use of peripheral nodes in routing.

For the OLSR, Figure 6.3 shows a reduction of load in central nodes (within a radius of 50 m) of around 55%, and an increase in the load of nodes further from the center (within a radius of 200 m) of around 300%.

As for AODV (Figure 6.4), the load of central nodes within a radius of 30 m decreased by 44% and that of peripheral nodes (200 m from the center) increased by a ratio of up to 1:4 when compared with the initial load.

Our load-sharing mechanism effectively made packets use paths further from the center of the network, reducing the load of central nodes and increasing that of peripheral nodes.

6.5.2. *Evaluation: end-to-end delay*

Figures 6.5 and 6.6 show the evolution of end-to-end delay as a function of mobility in the OLSR and the AODV. This delay is defined as the average time taken to transfer a data packet from a CBR source to a destination [LEE 05].

On the whole, we note a significant reduction in delay of up to one half in a network with low mobility for both protocols. By using longer but less-loaded paths, packets experience an increased transmission delay but spend less time in node buffers.

As transmission delays are considerably shorter than the time packets spend in node queues,[5] the overall effect is positive and the end-to-end delay is reduced.

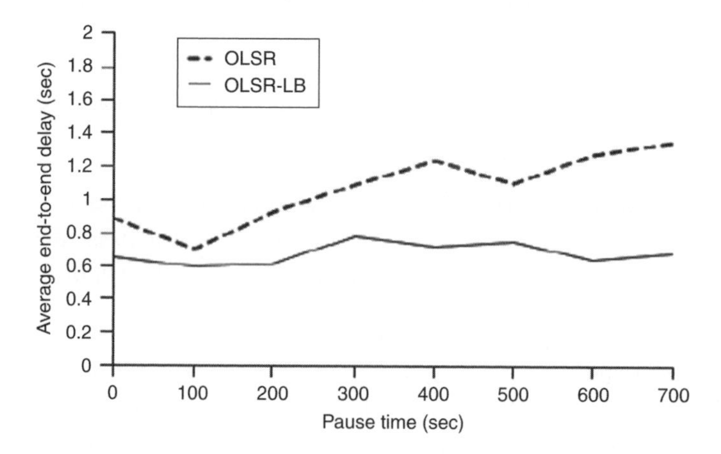

Figure 6.5. *Average end-to-end delay using OLSR*

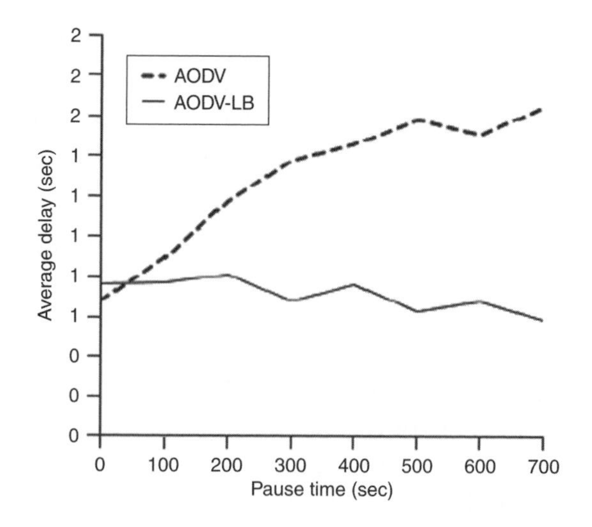

Figure 6.6. *Average end-to-end delay using AODV*

5 Transmission delay is measured in nanoseconds, whereas queuing time is measured in milliseconds, so the relationship in size is of the order 1:10.

6.5.3. *Evaluation: packet delivery fraction*

In these simulations, we have attempted to visualize the impact of our load-balancing mechanism on the reliability of transmissions.

The parameter to evaluate in this case is packet delivery fraction.

This parameter is defined as the relationship between the number of data packets received by destinations and the number of data packets generated by sources [LEE 05].

Figures 6.7 and 6.8 show the evolution of the packet delivery fraction in relation to mobility, with and without load balancing. AODV and OLSR are known for high levels of reliability when compared with other routing protocols.

To give an example, AODV offers a PDR of more than 90 % for debits under 500 kbit/s [LEE 05].

However, as soon as the load increases, the reliability of AODV decreases considerably [LEE 05]. For the scenarios envisaged in our simulation, we experienced PDRs of around 75%.

Thanks to our load-balancing mechanism, reliability was considerably increased, producing PDR values between 77 and 92%.

Once again, this improvement is due to the use of less busy routes. If packets travel through nodes with low loads, they stand less chance of arriving at a saturated queue and thus the probability of rejection is lower.

Note that this reliability is better at lower mobility rates in both cases, as mobility creates problems of link breakage, increased interference, and greater risk of collisions.

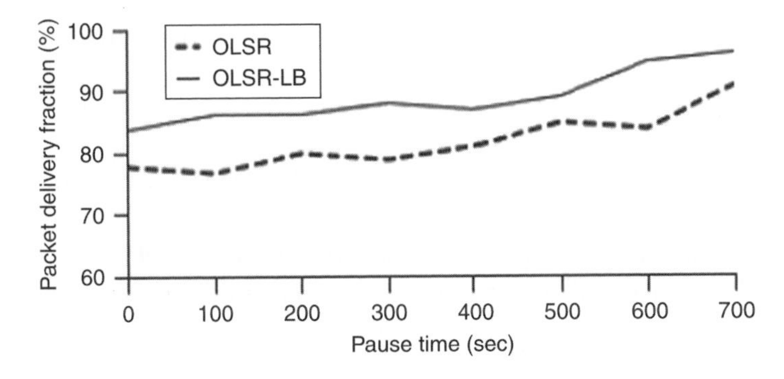

Figure 6.7. *PDR in OLSR*

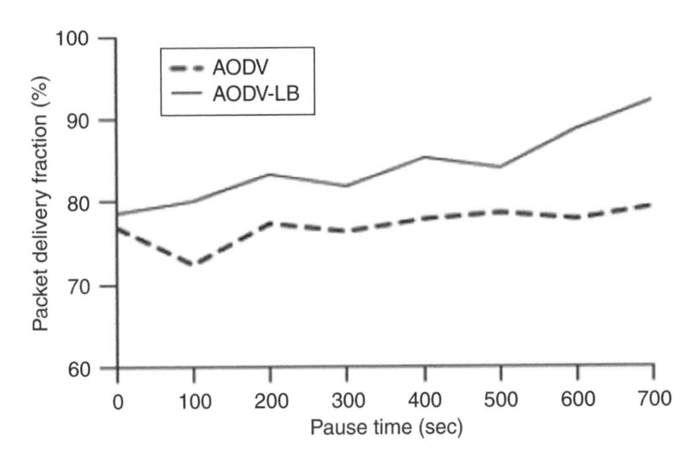

Figure 6.8. *PDR in AODV*

6.6. Conclusion

Starting from the observation that the spatial distribution of load in MANETs was extremely disparate and that load was the greatest at the center of the network, we have proposed new routing metrics for existing protocols with the aim of moving load away from the center of a network.

To this end, we separated proactive and reactive approaches and provided a characterization of the central nodes of the network for each.

We implemented the proposed mechanisms in implementations of the AODV and OLSR routing protocols using the ns-2 simulator [NS 07].

The results of our simulations have shown a definite improvement in the performance of the protocols in terms of end-to-end delay and packet delivery rate, in addition to the desired result of offering a more balanced distribution of traffic across the network.

Chapter 7

Energy Optimization in Routing Protocols

7.1. Introduction

In wireless networks, and ad hoc networks in particular, terminals are powered by batteries with limited lifespan. In networks with infrastructures, however, base stations do not have any limitations as far as energy is concerned; energy constraints are therefore only a problem for mobile units.

For this reason, strategies adopted by these networks are based on energy consumption by the base stations to maximize the lifespan of terminals. This approach does not work for ad hoc networks that do not have a preexisting infrastructure.

Furthermore, in networks with infrastructure, terminals operate independently of each other, using base stations to communicate between themselves. In this case, energy management for a node only concerns its own local applications. This is not the case in ad hoc networks, where nodes are linked and interdependent and cooperate in the routing process.

It is therefore important to develop strategies to maximize the lifespan of stations and thus of the whole network.

7.2. Energy optimization techniques

The main aim of ad hoc networks is to maintain a maximum number of active mobile terminals. In other words, access to information must be guaranteed wherever and whenever it is required. Energy supply problems are a major hindrance in attaining this aim. The energy constraints of ad hoc networks are essentially due to the use of batteries, which have very limited capacity.

7.2.1. *Energy consumption in ad hoc networks*

Energy consumption in ad hoc networks can occur in different layers of the reference model:

– *The physical layer*. Consumption due to the alimentation of different physical resources such as the microprocessor, display memory, and CPU.

– *The data liaison layer*. Energy used by collision and congestion control protocols.

– *The network layer*. Energy consumption by routing protocols maintaining radio transmissions, whether sending, receiving, or even waiting for a packet. For any given node, the total energy consumption of the radio interface can be modeled by a linear equation:

$$\text{total energy consumption (in Joules)} = (Pt \times Tt) + (Pr \times Tr) + (Pi \times Ti) \quad [7.1]$$

where Pt, Pr, and Pi, are the power used by a node sending, receiving, and waiting over time Tt, Tr, and Ti, respectively. Note that sending is more costly than receiving and that waiting is the least costly of the three. For any given communication, the energy consumption connected with one packet is the sum of the power required by the source to send the packet, the cost of reception by the destination, and the energy consumed during its passage through various intermediate nodes on the way to the destination.

– *The application layer*. At application layer level, battery capacity is used up by applications, particularly those responsible for the compression or encryption of data.

7.2.2. *Energy minimization methods for ad hoc networks*

As battery usage can be linked to the different layers of the reference model, a reduction in overall energy consumption can be obtained by optimizing one or more of these layers.

7.2.2.1. *The physical layer*

The physical layer plays an important role in establishing and maintaining communications between two nodes. The physical layer is responsible for the modulation and coding of data exchanged over a physical link between two radio interfaces, so that the intended receiver may decode the information correctly even in case of interference over the radio channel.

Research on energy consumption by the physical layer has concentrated on the material used. However, the most costly action for a node at this level is the transmission of a packet across the radio channel.

Several solutions have been proposed to deal with this problem, including the implementation of a power control policy. In ad hoc networks, the transmission power optimization approach produces a significant reduction in energy consumption. This approach has a major impact on battery life and the overall capacity of the network in terms of the quantity of traffic transmitted.

This solution is provided by implementing the variable power transmission protocol. To conserve energy, the protocol dynamically calculates the distance from source to target node. In this way, the exact level of power required for the transmission can be determined.

Using this method, the power usage of a source node for transmission is not constant, but depends on the distance between the source and the destination. Consequently, we can conclude that transmission power *Pt* is proportional to the distance between the two communicating nodes. Figure 7.1 shows that a node *A* must transmit packets to node *B* at a power of 2 mW and to node *C* at 20 mW.

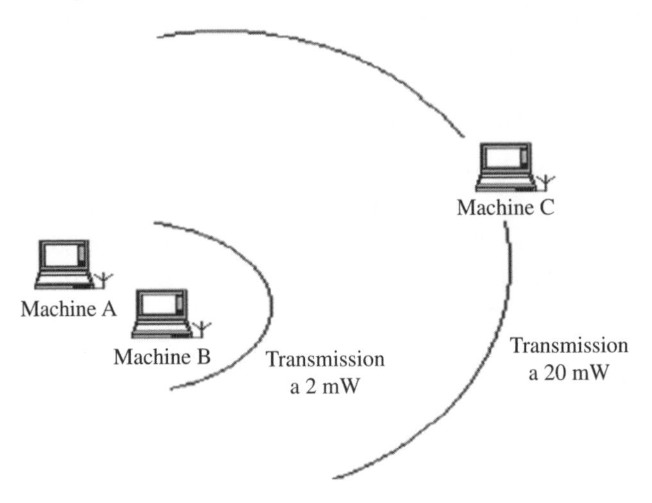

Figure 7.1. *Power transmission control*

7.2.2.2. *The data liaison layer*

The data liaison layer has two sublayers:

– the *medium access control* (MAC) sublayer; and

– the *logic link control* (LLC) sublayer.

7.2.2.2.1. The MAC layer

This layer is responsible for interfacing with the physical layer and defines the manner in which radio channels, shared between all of the mobile nodes, will be allocated.

This layer also allows collisions to be eliminated as far as possible. Every time a collision occurs, data must be retransmitted, leading to superfluous energy consumption.

Several MAC protocols exist, of which power aware multiple access protocol with signaling (PAMAS) provides the most significant reduction in energy consumption.

The PAMAS protocol is based on the multiple access collision avoidance (MACA) protocol, which is able to cope with hidden station problems. The PAMAS protocol is distinct from the MACA protocol in that it uses two separate channels, one for signalization, transporting RTS/CTS control messages, and another for data transmission.

The main advantage of PAMAS over MACA is the conservation of node battery power. This is achieved through optimization of the MAC layer, allowing stations to switch off their radio interface to reduce consumption every time the interface is unusable.

The protocol is based on the fact that a station cannot receive a message while listening to another communication, nor can it emit messages if a neighbor is receiving a message to avoid interference.

Any node in the network may switch off its radio interface, or "sleep," in the following cases:

– The node does not wish to send a message and at least one neighbor is transmitting to another station, as in this case, even if packets are sent to the node, it will be unable to receive them correctly.

– The node wishes to send and at least one neighbor is receiving, as, if it sends the message, it may disturb the neighbor.

– The node wishes to send and all of its neighbors are sending, as, in this case, none of the neighbors is available to transmit the data.

In this way, the PAMAS protocol provides energy savings of between 40 and 70%.

7.2.2.2.2. The LLC layer

The LLC layer deals with error control. For mobile ad hoc networks (MANETs), the use of error control methods such as ARQ and FEC, used in wired networks, creates very high-energy consumption due to retransmissions and the extra traffic load necessary for error correction.

One solution proposed for such situations is based on the introduction of a sensing protocol that allows data transmission to be slowed down when the radio channel is damaged.

The ARQ protocol is used normally unless the sender detects an error in the data or over the control channel due to the nonreception of an acknowledgment (ACK) message. When this happens, the protocol passes into sensing mode, where a sensing packet is sent every time slot *t*. The sensing packet contains only a header in order to minimize energy consumption. Once ACK has been received correctly, the protocol reverts to normal mode. Data transmission then resumes from the point at which it was interrupted.

7.2.2.3. *The network layer*

The largest class of energy-saving protocols applies to the network layer. Several research projects have been carried out in this domain, both to study the energy aspect of existing protocols and to develop the new routing protocols based on criteria including energy constraints.

The energy consumption of the network layer can be limited by reducing the charge needed to relay routes. To achieve these objectives, new metrics have been proposed. These metrics must respond to the following demands:

– reduce energy consumption as far as possible;

– maximize network lifespan;

– reduce power variation between nodes to a minimum;

– reduce costs per packet to a minimum; and

– reduce the maximum cost of a node as far as possible.

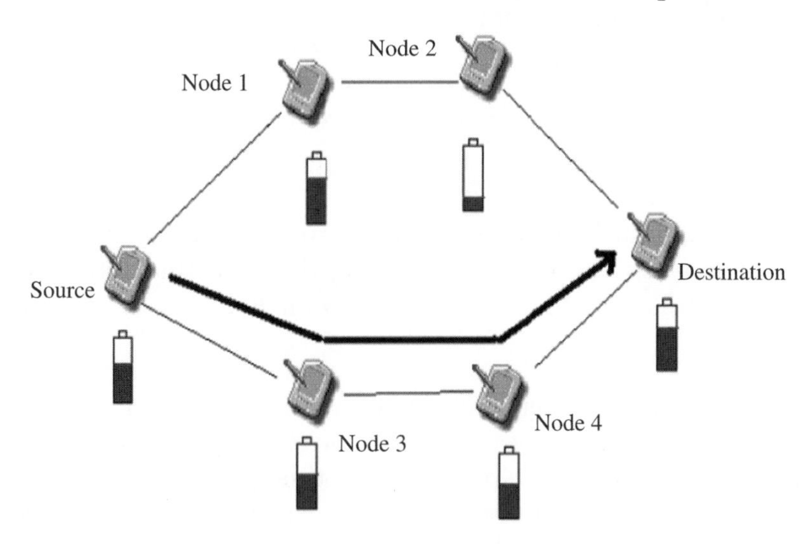

Figure 7.2. *Selection of routes with low-energy consumption*

In Figure 7.2, two paths from the source exist which allow data to be sent to the destination:

– source, node 1, node 2, destination and

– source, node 3, node 4, destination.

If we consider the minimization of energy usage and the maximization of the lifespan of the network as principal criteria in route selection, the second route is preferable as node 2 has the lowest power level.

To reduce energy consumption, we must improve pathfinding techniques and reduce the number of control messages sent as far as possible.

The basic idea that would allow energy resources to be optimized in the routing process is to route packets according to one or more criteria which impact upon energy consumption. These criteria are as follows:

– global minimization;

– local minimization; and

– modulation of transmission power.

7.2.2.3.1. Global minimization

This approach consists of minimizing the energy required to reach a destination while finding the path that consumes the least power. This approach is well adapted to reactive protocols.

Among the various solutions proposed, we find the idea of giving a weighting to each link according to the energy needed to transmit a packet over this path. In this case, routing is carried out with preference given to paths with low total weightings (the aggregate of the links making up the path).

Another solution would be to take the battery state of each terminal into account and to prefer those terminals with most remaining battery life to increase the lifespan of the network. In this case, each node estimates its remaining battery time based on previous activity. During the pathfinding procedure, the routing protocol chooses the route with the highest estimated lifespan, i.e. avoiding routes containing nodes with low remaining battery life.

7.2.2.3.2. Local minimization

This approach aims to increase the lifespan of the system using the willingness of nodes to participate (or not) in route selection procedures and in relaying traffic.

Based on its energy levels, a node may refuse to relay packets for which it is not the destination to conserve battery life.

7.2.2.3.3. Transmission power modulation

This last approach is based on the optimization of transmission power. Low-power transmission increases the capacity of the network while reducing energy consumption.

In this way, a route with more hops may be considered more efficient than a route with fewer hops.

However, we know that the risk of route failure increases in routes with higher numbers of hops, increasing the risk of retransmissions.

7.2.2.4. *The transport layer*

The transport layer provides an end-to-end service. The transport protocol most used in fixed networks is TCP.

TCP is not adapted to use in wireless networks as it generally uses a large number of retransmissions, and its congestion control mechanisms in case of losses or errors consume battery life unnecessarily.

It would therefore be interesting to develop transport protocols adapted to a wireless environment that would enable energy savings. Several suggestions have already been made in this domain, including the Reno and New Reno transport protocols.

7.2.2.5. *The application layer*

Energy optimization in the application layer is a major research domain. The solutions proposed include the principle of data encryption, to cite just one example.

The basic idea is to reduce the number of bits transmitted to reduce energy consumption while maintaining an acceptable level of visual quality.

7.3. Energy minimizing routing models in ad hoc networks

Four different power models have been defined with the aim of reducing energy consumption and maintaining battery life:

– minimum total transmission power routing (MTPR);

– minimum battery cost routing (MBCR);

– minimum–maximum cost routing (MMBCR); and

– conditional max–min battery capacity routing (CMMBCR).

7.3.1. *Minimum total transmission power routing*

In this routing model, for the transmission of data from node N_i to node N_j to succeed, the signal-to-noise ratio ($\mathrm{SNR}_{i\to j}$), for node N_j, must correspond to equation [7.2]:

$$\mathrm{SNR}_{i\to j} = \frac{P_i G_{ij}}{\sum_{k\neq i} P_k G_{kj} + \eta_j} \geq \Psi_j \qquad [7.2]$$

where $\mathrm{SNR}_{i\to j}$ is the SNR received by node N_j, P_i the transmission power of node N_i, G_{ij} the gain between nodes N_i and N_j, η_j the Johnson–Nyquist noise from node N_j, and ψ_j the protection threshold.

Minimum transmission power depends on the level of interference noise, the distance between the nodes and the bit error rate level.

The total transmission power from source to destination ($N_0 \rightarrow N_1 \rightarrow \ldots N_d$) is expressed by equation [7.3]:

$$P_j = \sum_{i=0}^{d_j - 1} P(N_i, N_{i+1}) \qquad [7.3]$$

where P_j is the total transmission power from source to destination, N_0 the source node, N_d the destination, and d_j the number of nodes on a path J (from the source to the destination).

The objective function for minimization is equation [7.4]:

$$P_k = \min_{j \in \Re} P_j \qquad [7.4]$$

where P_k is the minimum total transmission power from source to destination, P_j the total transmission power for a path J from the source to the destination, and $\Re$ = {all possible routes to the destination}.

The Dijkstra or Bellman–Ford shortest-path routing algorithm is used to resolve equation [7.4]. Algorithms of this kind present certain disadvantages:

– increased end-to-end delay, in cases where the path contains one or more hops; and

– mobility makes the path unstable.

To resolve these problems, transmission power and reception power must be taken into account, hence:

$$C_{i,j} = P_{\text{transmit}}(N_i, N_j) + P_{\text{transceiver}}(N_j) + Cost(N_j) \qquad [7.5]$$

where $C_{i,j}$ is the minimum energy requirement for the establishment of a route from node N_i towards the destination (via node N_j), P_{transmit} (N_i, N_j) the transmission power from node N_i to its neighbor N_j, $P_{\text{transceiver}}$ (N_j) the power necessary for reception by node N_j, and $Cost(N_j)$ the minimum energy necessary to establish a route from node N_j to the destination.

The disadvantages of this model are that it does not reflect the individual lifespan of each node and that each node must regularly, on waking up, rediscover its neighbors.

7.3.2. *Minimum battery cost routing*

Total transmission power is an important metric as it gives the lifespan of a mobile node. However, this metric has one critical shortcoming.

Although it allows global energy consumption to be reduced, this metric does not take the lifespan of individual nodes into consideration. If all routes with minimum total transmission power pass through the same intermediary node, the battery of this node will be drained rapidly and the node will then disappear from the network.

Consequently, it would be useful to take the remaining battery capacity of each node into account as a life-expectancy metric.

The cost function of a battery can be expressed in equation [7.6]:

$$F_i(C_i^t) = \frac{1}{C_i(t)} \quad [7.6]$$

where $F_i(C_i^t)$ is the cost function of the battery of node N_i at the instant t and $C_i(t)$ gives the capacity of the same battery at the same instant.

From equation [7.6], we see that as battery capacity declines, so the cost function of node N_i increases. The battery cost for a route J, noted R_j, is given in equation [7.7]:

$$R_j = \sum_{i=0}^{d_j-1} F_i(C_i^t) \qquad [7.7]$$

where R_j is the battery cost of a route J from the source to the destination and d_j the number of nodes participating in route J.

To conserve as much battery capacity as possible, we must choose the route I with the lowest battery cost consumption:

$$R_i = \min\left\{\frac{R_j}{j \in A}\right\} \qquad [7.8]$$

where R_i is the minimum battery cost of routes from the source to the destination, i.e. the cost of route I, and A is all possible routes (J) from the source node to the destination.

This model presents the following advantages:

– nodes will not be overused as battery capacity is used by the routing protocol; and

– in cases where nodes all have the same battery capacity, the shortest path (i.e. with the least hops) will be used.

The disadvantage of this energy optimization model is that only the sum of the best cost functions is taken into consideration, so a route containing nodes with low battery could still be selected.

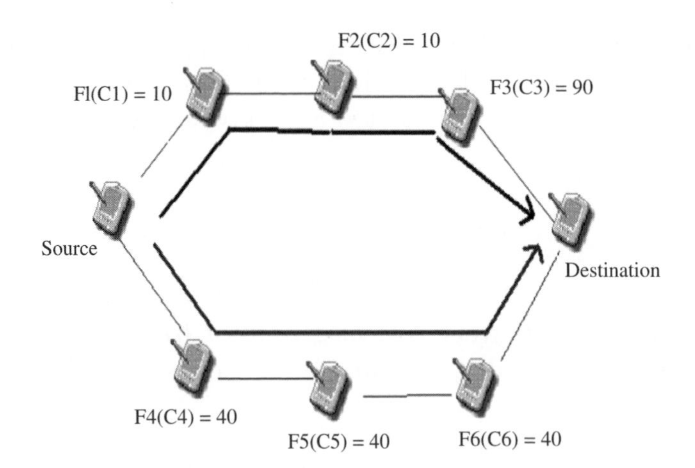

Figure 7.3. *Route selection principle using the MBCR model*

From Figure 7.3, we see that two possible routes exist from the source to the destination:

– source, node 1, node 2, node 3, destination; and

– source, node 4, node 5, node 6, destination.

We note that the battery cost of route 1 is lower than that of route 2, so, according to MBCR, route 1 will be chosen. This choice may reduce the lifespan of node 3, an undesirable effect of this decision.

7.3.3. *Min–max cost routing*

To ensure that no node is overused, the cost function can be modified. Equation [7.7] can then be rewritten as follows:

$$R_j = \max_{i \in \text{route}_j} F_i(C_i^t) \qquad [7.9]$$

where R_j is the battery cost of a route J from the source to the destination, $F_i(C_i^t)$ the battery cost function of node N_i at time t, and $route_j$ a route J from the source to the destination.

For the route with the lowest battery cost consumption, I, we have:

$$R_i = \min\left\{\frac{R_j}{j \in A}\right\} \qquad [7.10]$$

where R_i is the minimum battery cost of routes from the source to the destination, i.e. the cost of route I, and A = {all possible routes (J) from the source node to the destination}.

As this metric always tries to avoid using paths where one or more nodes has low-battery capacity, nodes on established paths will be used more in the case of the MMBCR model than in the MBCR.

The disadvantage of this model is that there is no guarantee that the route with the lowest energy requirements will be chosen under all circumstances. It is therefore possible that more power than strictly necessary will be used in transferring useful data from source to destination, reducing the lifespan of all the nodes involved.

7.3.4. *Conditional min–max battery capacity routing*

We have seen that the MTPR and MMBCR models are unable to produce a compromise between maximizing the lifespan of each individual node and prolonged battery use.

In this model, remaining battery capacity is used as a routing decision metric instead of the cost function. When nodes between the source and the destination have a capacity above a certain threshold, then MTPR is used. If battery capacity is lower than the threshold, the route in which the nodes have the lowest remaining battery capacity must be avoided to prolong the life of the network. The battery capacity of a route J at time t is given by equation [7.11]:

$$R_j^c = \min_{i \in \text{route}_j} C_i(t) \qquad [7.11]$$

where R_c^j is the battery capacity of a route J at time t, $C_i(t)$ is the battery capacity of node N_i at time t, and $route_j$ is a route J from the source to the destination.

Let us suppose that B is the group of all possible routes between the source and the destination at time t, which satisfies equation [7.12]:

$$R_j^c \geq r, (j \in B) \qquad [7.12]$$

where r is the threshold, or protection margin, and B = {all possible routes J between the source and the destination at time t so that $R_j^c \geq r$}.

We note that the performance of the CMMBCR model varies depending on the choice of threshold r.

Let A be the group of all possible routes between the source and the destination at time t. Two situations are possible at the moment when a route is needed between a source and a destination:

– If $A \cap B \neq \varnothing$, then all nodes in the network have battery capacity above the threshold. In this case, the MTPR model may be used to find the best route.

– If this is not the case, we must choose the route with maximum battery capacity, I:

$$R_i^c = \max \left\{ \frac{R_j^c}{j \in A} \right\} \qquad [7.13]$$

During the data transfer, this model offers the opportunity to introduce two new thresholds to carry out controls on the state of the battery of each node:

– selective victim search zone (SVSZ); and

– forced victim search zone (FVSZ).

7.3.4.1. *SVSZ*

SVSZ is used to indicate that a working node has low-battery capacity. If the route using the node with low remaining energy is still in use, the node enters the FVSZ and may die.

In this way, a certain path must be chosen to send packets in the network. This situation is illustrated in Figure 7.4.

Figure 7.4 shows the method of route formation between sources and destinations. We note that node *p* carries the most load and its battery is the most used. It is possible that thus node may enter the SVSZ.

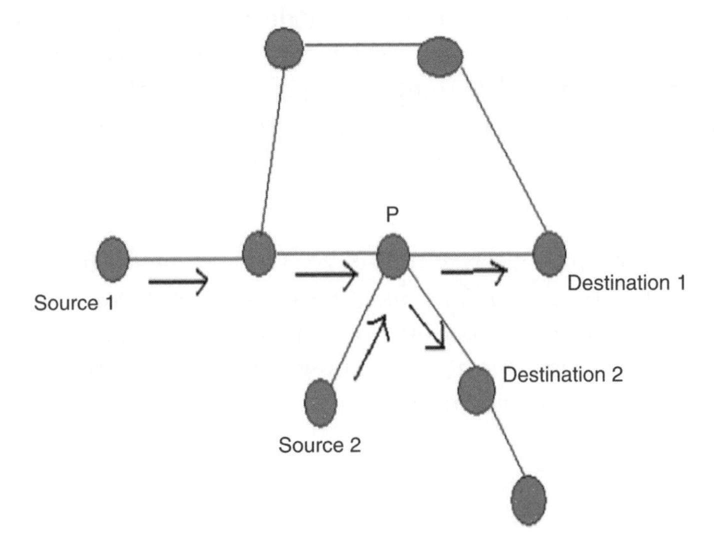

Figure 7.4. *Principle of route formation between sources and destinations*

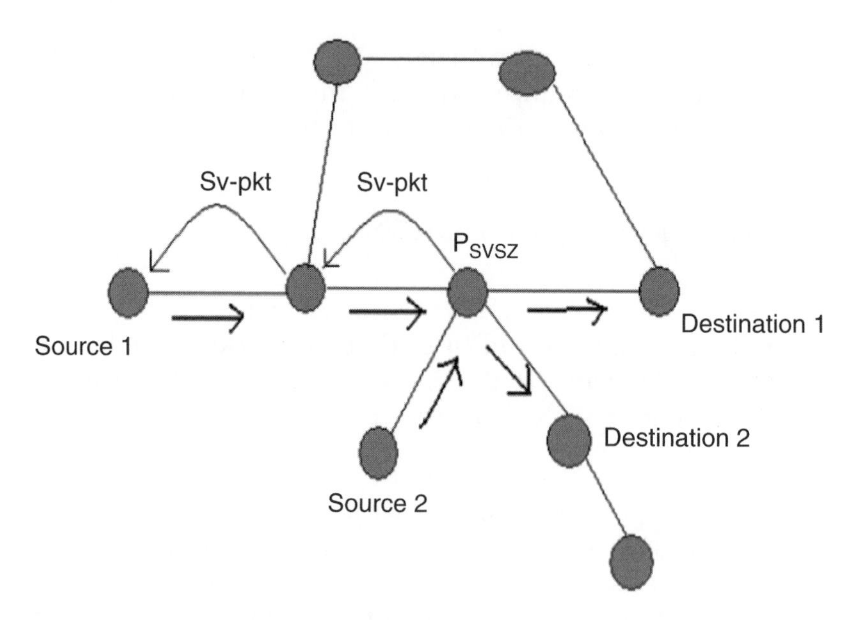

Figure 7.5. *Sending Sv-pkt packets to source 1*

As Figure 7.5 demonstrates, node *p* must send a Sv-pkt packet to the source for which the route consumes most energy (s1 to d1 or s2 to d2). Supposing that the transfer of data from source s1 to destination d1 consumes the most energy, then node *p* must choose source s1 as the victim, then send a Sv-pkt.

On receiving a Sv-pkt, source s1 starts a new route discovery process. In this case, node *p* does not participate in the pathfinding operation and does not send the route request message. Figure 7.6 shows the new route between source s1 and destination d1.

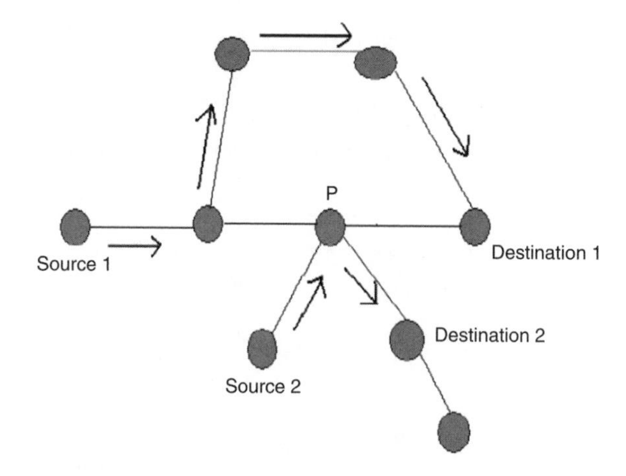

Figure 7.6. *New route from source 1 to destination 1*

7.3.4.2. *FVSZ*

As Figure 7.7 shows, node *p* participated in the transfer of packets from s2 to d2 and continues to do this even after entering the SVSZ state. This makes it highly probable that node *p* will enter the FVSZ state; however, until this happens, the node must continue to transmit packets.

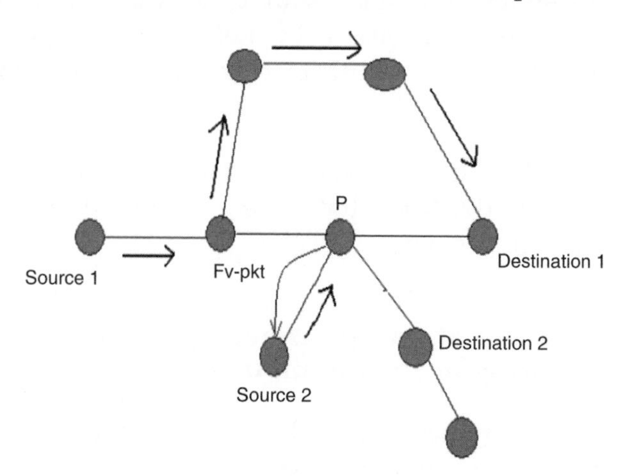

Figure 7.7. *Sending FV packets to source 2*

As the capacity of its battery becomes critical, node *p* sends a FV-pkt packet to source s2 so that a new route to destination d2 may be found. This is illustrated in Figure 7.7. Node *p* will then refuse to transmit packets sent by source s2.

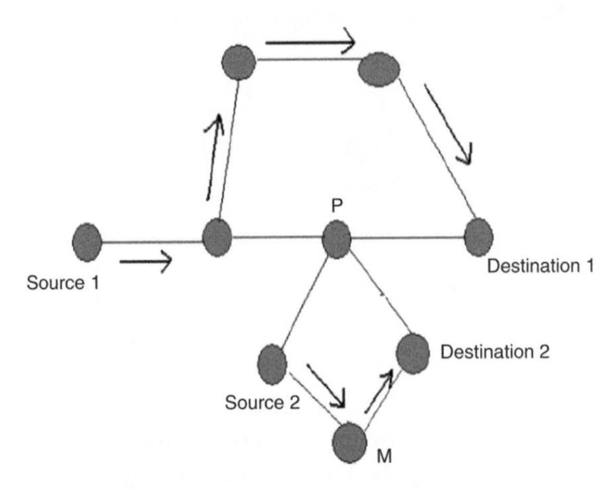

Figure 7.8. *Establishment of new routes*

After receiving a Fv-pkt, source s2 was unable to find another route to destination d2. Let us now suppose that node *M* belongs to the zone of coverage of both nodes s2 and d2.

As Figure 7.8 shows, node *M* is used to establish a new route from s2 to d2. Note that node *p* does not participate in the discovery of this new route.

7.4. Comparison of energy consumption for an ad hoc network routing protocols simulated in ns-2

The list of protocols already implemented in the ns-2 simulator includes dynamic source routing (DSR), ad hoc on-demand distance-vector (AODV), destination-sequenced distance-vector (DSDV), and TORA.

In what follows, we shall compare these four protocols with emphasis on the energy consumption of each.

7.4.1. *The energy consumption model*

According to the specifications of the NIC model (bandwidth = 2 mbps), energy consumption varies from 230 mA for receiving to 330 mA for transmission, using voltages of 3.3 or 5.0 V.

The following equations represent the energy used (in Joules) in transmitting (equation [7.14]) or receiving (equation [7.15]). Packet size is given in bits.

– for transmission:

$$\text{Energytx} = \text{txPower} \times (\text{packetsize/debit}) \qquad [7.14]$$

– receiving:

$$\text{Energyrx} = \text{rxPower} \times (\text{packetsize/debit}) \qquad [7.15]$$

The power used in transmitting and receiving is given by the following equations:

$$\text{txPower} = 5 \times 0.33 = 1.65 \text{ (Watts)} \qquad [7.16]$$

$$\text{rxPower} = 5 \times 0.23 = 1.15 \text{ (Watts)} \qquad [7.17]$$

Consequently, equations [7.14] and [7.15] become:

$$\text{Energytx} = \frac{330 \times 5 \times \text{packetsize}}{2 \times 106} \qquad [7.18]$$

$$\text{Energytx} = \frac{230 \times 5 \times \text{packetsize}}{2 \times 106} \qquad [7.19]$$

7.4.2. *Simulation method*

Our aim is to measure, then compare the behavior of the four routing protocols cited above in terms of energy consumption.

First, we need to choose the most representative parameters that serve our needs, then define and simulate a basic scenario. We can then simulate and evaluate different scenarios by changing these parameters.

The five parameters chosen for the simulations are as follows:

– simulation length,

– node mobility model,

– number of traffic sources,

– number of mobile nodes,

– dimensions of mobility zone, and

– traffic model.

For these simulations, we shall use the random waypoint mobility model. Node movement is characterized by two factors: maximum velocity and pause time.

During the simulation, each node begins to move from its initial position following a random path within the simulation zone. All traffic sources used in the simulations produce constant bit rate (CBR) traffic. The structure of this traffic is defined by changing two factors: sending rate and packet size.

7.4.3. *Simulation results*

As a basic scenario, we have chosen a MANET of 25 mobile nodes scattered at random over a surface of

500 × 500 m. Nodes move at a maximum velocity of 15 m/s with a pause time of 0 s.

The number of traffic sources is set at 10, each producing CBR traffic at a rate of around four packets per second with packet size of 512 bits. Each simulation will last approximately 600 s.

7.4.3.1. *Basic scenario simulation*

The basic scenario simulation provides an idea of the behavior of each protocol depending on simulation time.

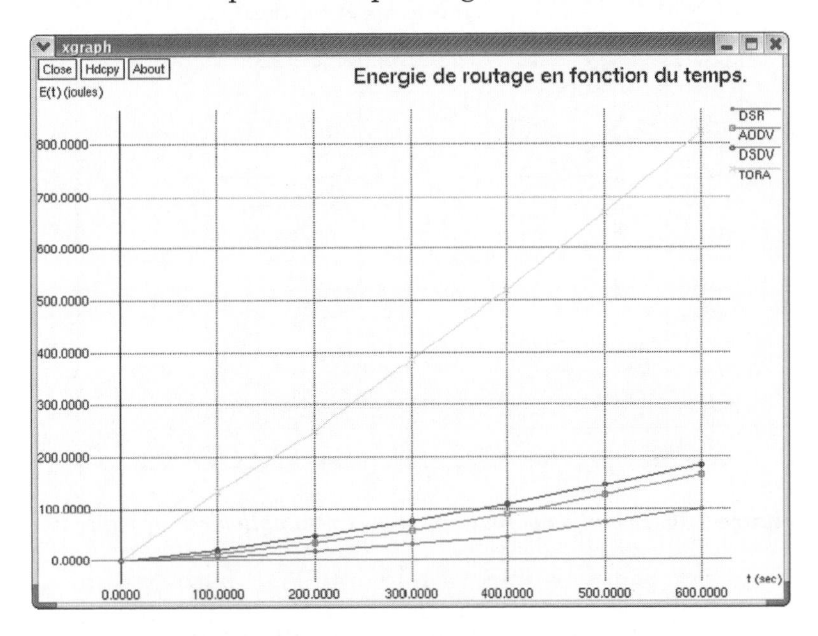

Figure 7.9. *Evolution of routing energy consumption versus simulation time*

From Figure 7.9, we see that the DSR protocol offers the best results, whereas TORA performs least well. The reactive DSR protocol performs better than the AODV in this respect as it is based on the source routing technique, whereas AODV uses the node-by-node routing principle.

We also note that the behavior of the proactive protocol DSDV is more like that of the AODV than that of the TORA.

7.4.3.2. *Pause time variation*

In this section, we shall explore the effect of varying mobility models in the basic scenario. We shall change the pause time and take measurements at pause times of 0 (nodes in continual movement), 100, 200, 300, 400, and 500 s and 600 s (static nodes).

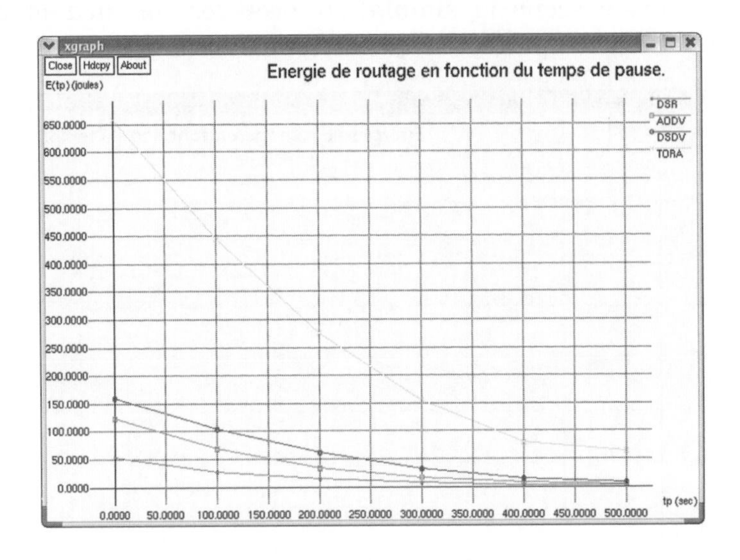

Figure 7.10. *Evolution of routing energy consumption versus pause time*

Figure 7.10 shows the energy consumption level of each routing protocol contingent on pause time. DSR offers the best results and TORA the weakest. As a general rule, we notice that for all four protocols (DSR, AODV, DSDV and TORA), energy consumption decreases as node movement decreases.

Of the reactive protocols under inspection, TORA presents the lowest results. This is largely due to the aggregation of the IMEP and TORA packets. DSR and AODV present the same behavior as the scenario converges to a static network.

7.4.3.3. *Variation of maximum node velocity*

Figure 7.11 shows the results obtained by changing the maximum velocity of node movement, using the following values: 0, 1, 5, 15, and 25 m/s. These values have been chosen to simulate the following real-world scenarios:

– a static network;

– a MANET for pedestrians;

– a MANET for a group of cyclists;

– a MANET for cars in an urban environment; and

– a MANET for cars on a motorway.

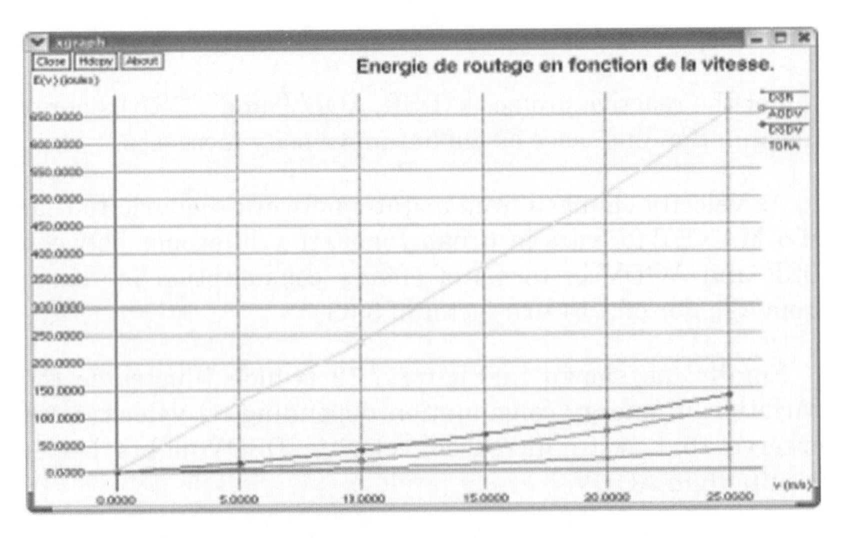

Figure 7.11. *Evolution of routing energy consumption versus maximum velocity*

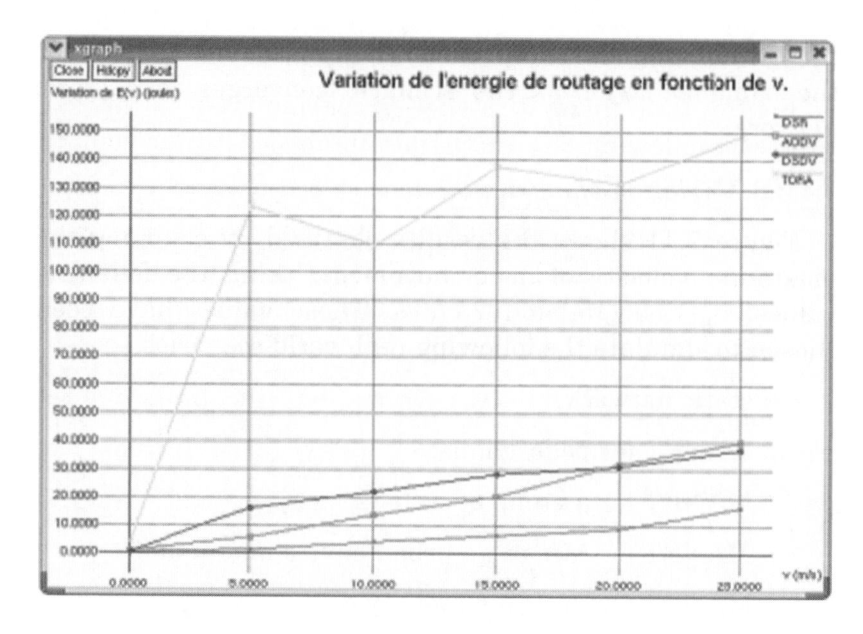

Figure 7.12. *Variation of routing energy consumption versus maximum velocity*

For the reactive protocols (DSR, AODV, and TORA), energy consumption increases as maximum velocity increases.

As velocity changes from a static network scenario to that of a MANET for cars in urban zones, the difference between DSR and AODV in terms of energy consumption increases from a factor of 2.14 to a factor of 3.63.

Finally, as shown in Figure 7.12 (which illustrates the variation in energy consumption depending on velocity), we observe that, with increased velocity, DSDV offers better results than AODV.

These results confirm the constant behavior of the DSDV with regard to energy consumption, which is almost constant for high-mobility simulations (Figure 7.11).

7.4.3.4. *Variation in number of traffic sources*

Figure 7.13 shows the energy consumption of the four protocols following change in the number of traffic sources, which was set at 10, 20, or 30.

For the DSR, AODV, and DSDV protocols, energy consumption increases at a faster rate than the number of sources as an increase in the number of traffic sources increases the number of routing packets used.

From Figure 7.13, we see that when the number of traffic sources increases from 10 to 20, energy usage increases by 214.1% for DSR, 302.18% for AODV, 159% for DSDV, and 120.18% for TORA.

As the number of traffic sources increases from 20 to 30, routing energy consumption increases by 177% for DSR, 182.3% for AODV, 118.6% for DSR, and 61.78% for TORA.

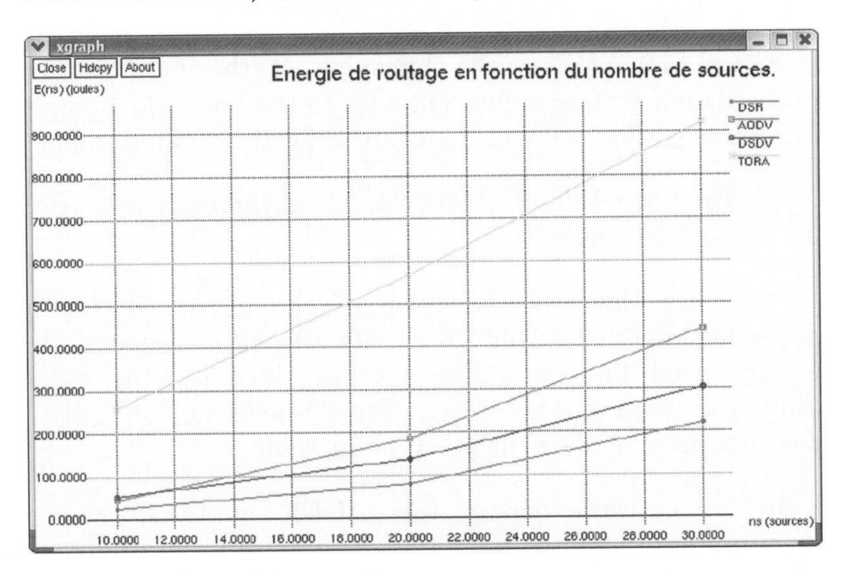

Figure 7.13. *Evolution in routing energy consumption depending on number of traffic sources*

Figure 7.14 shows the variation in energy consumption following variation in the number of traffic sources.

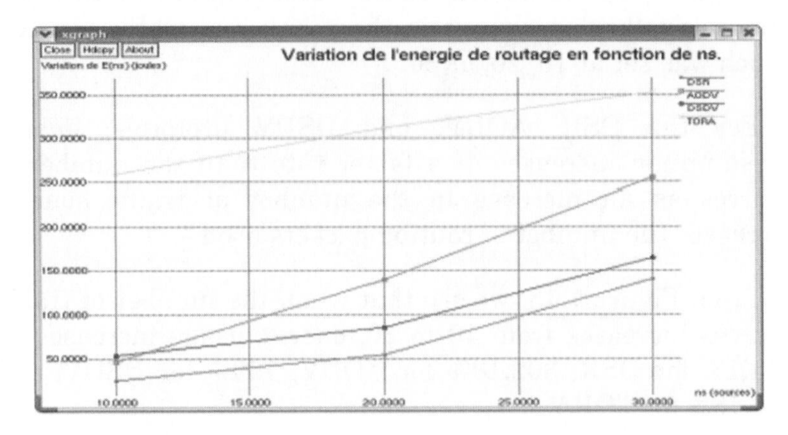

Figure 7.14. *Variation in routing energy consumption versus number of traffic sources*

7.4.3.5. *Variation in number of nodes*

Figure 7.15 shows the behavior of the four routing protocols when the number of nodes in the network changes. The same traffic load was maintained for these simulations.

We used MANETs with 10, 20, 25 (in the basic scenario), 30, 40, and 50 nodes.

We note that the behavior of the TORA is highly dependent on this factor. In a network with 50 nodes, the energy used by the TORA protocol increases by 833% compared with energy usage for 25-node network. This characteristic renders the protocol unstable.

In a network using the AODV protocol, energy consumption increases by 400% as we pass from a network with 25 nodes to a network with 50 nodes. This increase is mainly due to the route maintenance process. The increase of energy consumption in the DSDV is mainly a result of the propagation of the routing table between nodes.

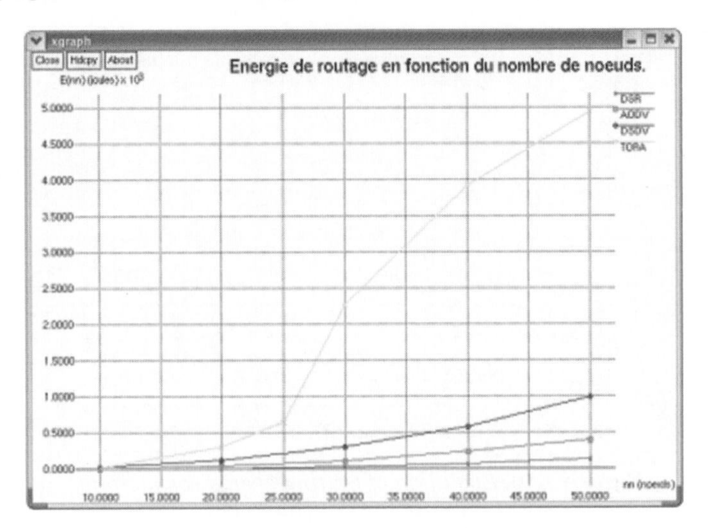

Figure 7.15. *Evolution of routing energy consumption contingent on number of nodes*

7.4.3.6. *Variation of test network surface*

Figure 7.16 shows the results of simulations where the surface dimensions of the network were varied. The following dimensions were used: 250 × 250 m, 250 × 500 m, 500 × 500 m, and 1000 × 500 m.

From Figure 7.17, we see that as the surface of the network grows, the energy consumption of the AODV increases more rapidly than the DSDV protocol (from a surface of 500 × 500 m).

In networks with large surface areas (500 × 500 m, for example), the proactive DSDV protocol performs better than the AODV protocol, and even better than the DSR protocol (from surface area of 445,000 m^2 onwards).

We also notice that the TORA protocol offers the weakest results when compared with the other three protocols.

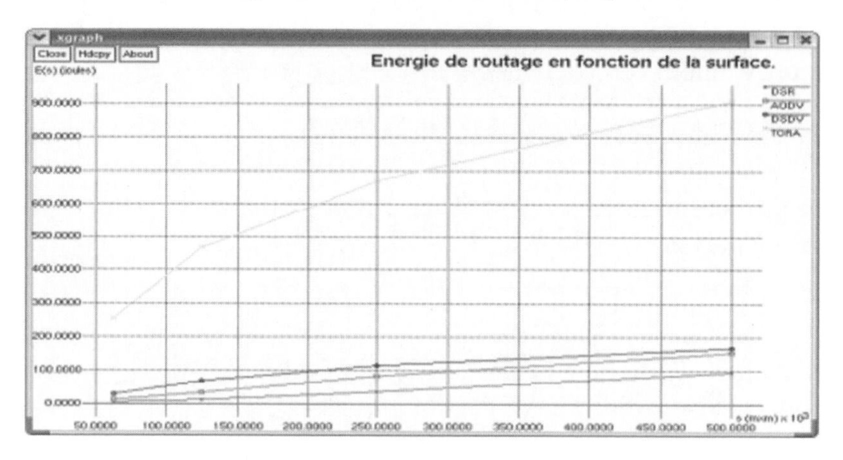

Figure 7.16. *Evolution of routing energy consumption versus network surface area*

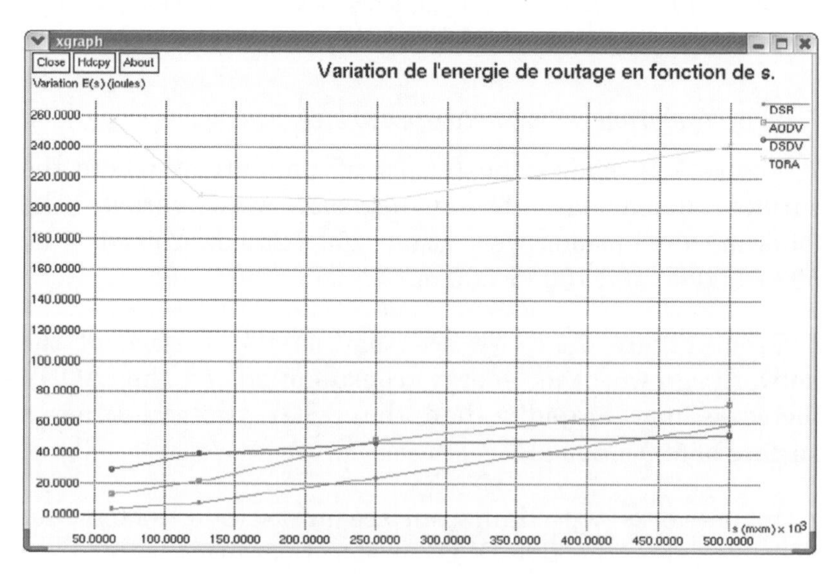

Figure 7.17. *Variation of routing energy consumption versus network surface area*

7.4.3.7. *Transfer debit variation*

Figure 7.18 shows the effects of increasing the packet emission rate. We note that almost all the protocols show improvements in their behavior.

This improvement is explained by the use of the routing table in the DSDV protocol and by the ability of the reactive protocols to recover routing information contained in previously received packets.

DSR and AODV offer the best results.

Figure 7.19 gives an idea of the way in which routing energy consumption evolves for the DSR, AODV, DSDV, and TORA protocols.

We can see that the variation in energy usage evolves in a very similar way in the DSR and the AODV.

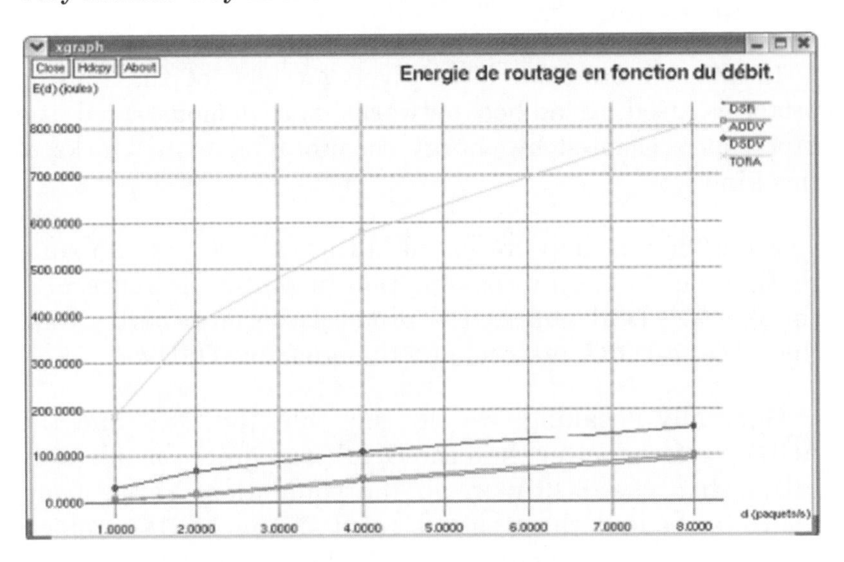

Figure 7.18. *Evolution in routing energy consumption versus transfer debit*

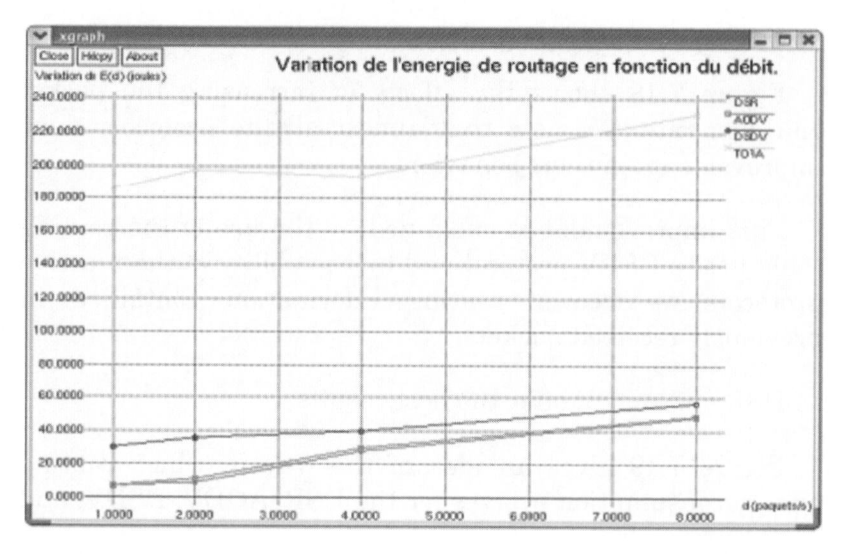

Figure 7.19. *Variation of routing energy consumption versus transfer debit*

7.5. Conclusion

In this chapter, we have given an idea of the routing protocols used in ad hoc networks and demonstrated the problems presented by energy consumption in networks of this kind.

From the research presented in this chapter, we can state that, as far as energy consumption in routing is concerned, the reactive DSR and AODV protocols perform better than the reactive TORA protocol and the proactive DSDV.

Generally speaking, we can say that the DSR and the AODV perform better than the DSDV and quite considerably better than the TORA; in all the simulations carried out, TORA presented the weakest results. The DSDV protocol presented constant behavior in all the simulated scenarios; this is due to the way in which it uses its routing table.

Chapter 8

Wi-Fi Access for Ad Hoc Networks

8.1. Introduction

The Institute of Electrical and Electronics Engineers (IEEE) norm 802.11 is an international standard describing the characteristics of a wireless local area network (WLAN).

The name "wireless fidelity (Wi-Fi)" initially referred to the name of the certification delivered by the Wireless Ethernet Compatibility Alliance, the body charged with maintaining interoperability between materials conforming to the 802.11 norm.

Through misuse of terminology (and for marketing reasons), the name of the norm has now become confused with the name of the certification. A so-called "Wi-Fi" network is, in reality, a network that follows the 802.11 norm.

Thanks to Wi-Fi, it is now possible to create broadband wireless local networks. In practice, Wi-Fi allows portable computers, desktops, PDAs, and even peripherals to connect to a high-speed link (from 11 Mbps in 802.11b to 54 Mbps in 802.11a/g) within a radius of several dozen meters indoors (generally, between 20 and 50 m).

In a more open environment, the range can attain several hundred meters, or even, in optimal circumstances, tens of kilometers.

8.2. Wi-Fi network structure

The 802.11 norm defines the lower layers of the OSI model for a wireless link using electromagnetic waves, i.e.

– the physical layer (sometimes noted PHY layer), proposing different types of information coding;

– the data link layer, made up of two sublayers:

- the logical link control (LLC) sublayer; and

- the medium access control (MAC) sublayer.

The physical layer defines the modulation of radio waves and the signalization characteristics for data transmission, whereas the data link layer controls the interface between the machine bus and the physical layer, an access method close to that used in the Ethernet standard and the rules for communication between different stations.

The 802.11 norm actually proposes three physical layers, defining alternative modes of transmission (Figure 8.1).

Data link layer	802.2 (LLC)			
	802.11 (MAC)			
Physical layer (PHY)	DSSS	FHSS	OFDM	Infrared

Figure 8.1. *802.11 layers*

Any transport protocol can be used in a Wi-Fi wireless network, as in an Ethernet network or the Internet.

8.2.1. *The physical layer*

The physical layer defines transmission techniques (modulation of radioelectric waves), encoding, and transmission signaling. Any sinusoidal electrical signal can vary its amplitude, frequency, and phase.

Any of these three parameters can be used to modify an electrical signal for coding purposes. Frequency and phase modulations are generally used together to improve performance.

The 802.11 physical norm proposes three types of transmission with frequency and phase modulations and an infrared transmission technique.

We shall only look at those types of frequency modulation transmission based on the spread spectrum technique or the orthogonality of carriers.

The spread spectrum technique, developed by the military, has found considerable favor as it effectively combats interference and allows several transmissions to coexist over the same wavelength.

The spread spectrum technique is thus the basis of normalization for several wireless norms.

Table 8.1 shows different spread spectrum techniques by the 802.11 technology. Most use a wavelength between 2.4 and 2.4835 GHz.

This wavelength is known as ISM (Industrial, Scientific, and Medical), and can be used without prior authorization.

Technology	Type of spread spectrum	Transmission frequency	Max. debit (Mbps)
IEEE 802.11	FHSS, DSS, and IR	ISM	2
IEEE 802.11b	DSSS	ISM	11
IEEE 802.11a	OFDM	5.15 to 5.35GHz and 5.725 to 5.825 GHz	54
IEEE 802.11g	DSSS and OFDM	ISM	54

Table 8.1. *Spread spectrum technique*

8.2.1.1. *Frequency hopping spread spectrum (FHSS)*

The idea behind FHSS is to share the information signal across a larger bandwidth to make interference and/or interception more difficult.

This technique modifies the frequency of the carrier by a sequence of hops, i.e. the sender changes frequency periodically following a preestablished sequence (Figure 8.2). The receiver is synchronized using tagged frames containing the hop sequence and the time between frequency changes.

In the 802.11 norm, the ISM wavelength, from 2.400 to 2.4835 GHz, is split into 79 channels of 1 MHz and hops occur every 300–400 ms. The sender and the receiver agree on one of the several possible hopping sequences.

The norm defines three groups of 26 possible sequences each, i.e. a total of 78 sequences.

The signals (data transformed by FHSS) are then modeled by a phase modulation such as Gaussian frequency shift keying. Debit reaches 1–2 Mbps.

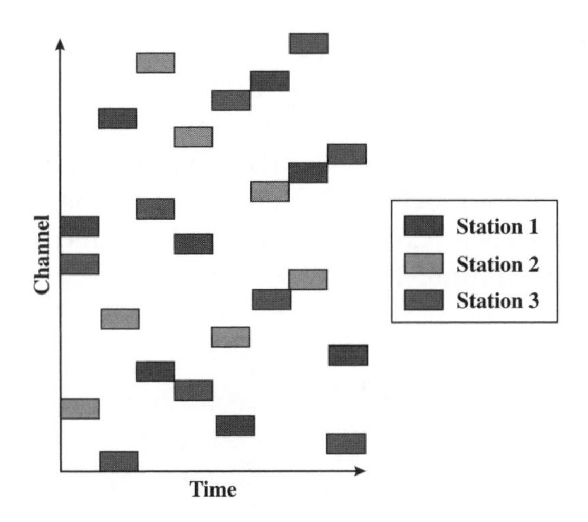

Figure 8.2. *The frequency hopping principle*

This technique was initially used by the military to encrypt transmissions, but the frequency sequences are now standardized and have therefore been divulged. The 802.11 norm uses this technique to deal with interference.

The technique presents several advantages as it offers signal immunity to different sorts of noise and prevents distortion caused by multipath propagation. Moreover, it can be used to hide signals, as only the receptor that knows the spreading code can retrieve the coded information.

Finally, this technique allows several users to exploit the same spread bandwidth with little or no interference. The principal disadvantage of FHSS is that the debit is limited to 2 Mbps due to the limitation of bandwidth over the channels to 1 MHz by the Federal Communications Commission.

8.2.1.2. *Direct sequence spread spectrum (DSSS)*

DSSS is another spreading technique available in the 802.11 norm.

Like FSSS, DSSS operates over the ISM wavelength. In this case, the bandwidth is divided into 14 channels of 20 MHz, each made up of four units of 5 MHz. As the bandwidth available is 83.5 MHz, the 14 channels cannot be placed without overlap. Channels are therefore spaced at 5 MHz distance, except for channel 14 that is spaced at 12 MHz from channel 13.

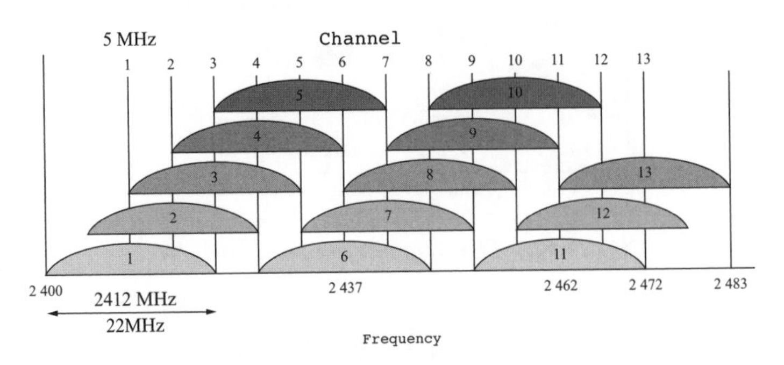

Figure 8.3. *Placement of transmission channels*

When a channel is selected, the signal spectrum occupies a space of 10 MHz either side of the node frequency. For this reason, we can only use three distinct sending channels on the same cell without the risk of interference (Figure 8.3).

The principle of DSSS involves coding each bit of the original signal over several bits of the transmitted signal using a spreading sequence, also known as "chipping" or "Barker's sequence", as shown in Figure 8.4. The signal is then transmitted over a channel of 20 MHz.

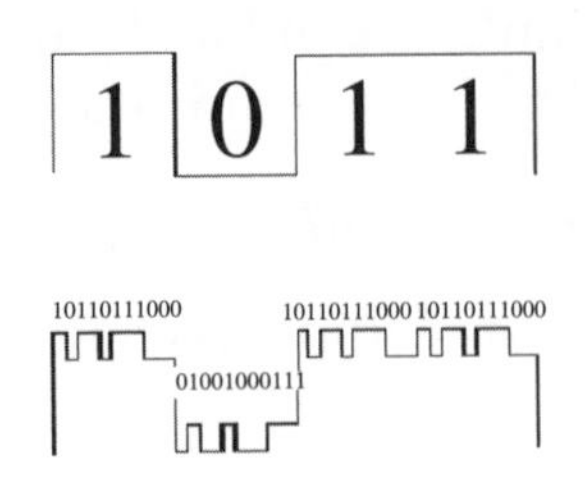

Figure 8.4. *Principle of direct sequence with Barker's sequence for Wi-Fi*

Thanks to DSSS, redundant information is transmitted over the radio channel. This allows error controls to be carried out and even error correction in transmissions.

Nevertheless, to attain an acceptable transmission rate of 11 Mbps, transmission must be carried out over a bandwidth of 22 MHz as, according to Shannon's theorem, sampling frequency must be at least the double of the signal to digitize.

Following this principle, in high rate-DSSS (HR-DSSS) a derivative of DSSS, bandwidth is divided into just 11 channels of 22 MHz.

To avoid interference between transmissions using DSSS, channels 1, 7, and 13 are isolated at 25 MHz from each other. Generally, HR-DSSS uses channels 1, 6, and 11 for this purpose, spaced out by 30 MHz.

The disadvantage of DSSS is that signal perturbations can occur if two peripherals are communicating at the same time through overlapping channels.

8.2.1.3. *Orthogonal frequency division multiplexing (OFDM)*

OFDM is a multicarrier modulation.

This technology was developed in 1960s, but is used considerably more nowadays as signal treatment techniques make it more feasible and more efficient.

The OFDM principle involves splitting the available bandwidth into slots, known as carriers or subcarriers, which are distinct channels for data transmission.

Carriers are distinguished from each other as the node of one carrier corresponds to zero amplitude for the adjacent carriers (orthogonality; Figure 8.5).

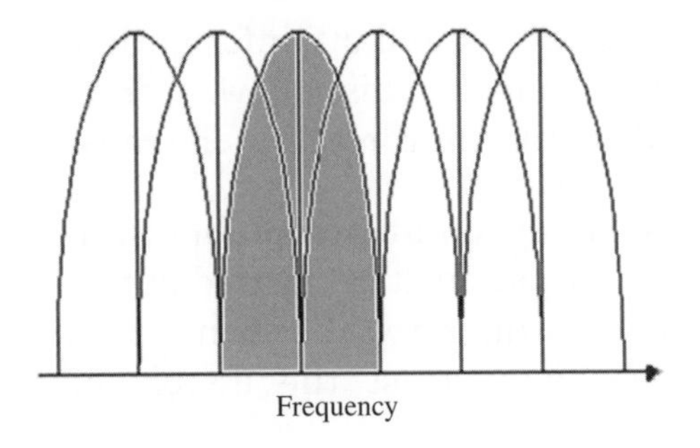

Figure 8.5. *OFDM principle*

To do this, the OFDM technique takes the coded signal on each subchannel and uses the inverse fast Fourier transform algorithm to generate a composite wave from the power of each subchannel. Receivers do the opposite, i.e. decompose the composite wave into signals over different channels using the fast Fourier transform algorithm.

A main advantage of the OFDM is its robustness when dealing with multiple communications and the fact that data transmission debit can reach 54 Mbps over the ISM bandwidth (Figure 8.6).

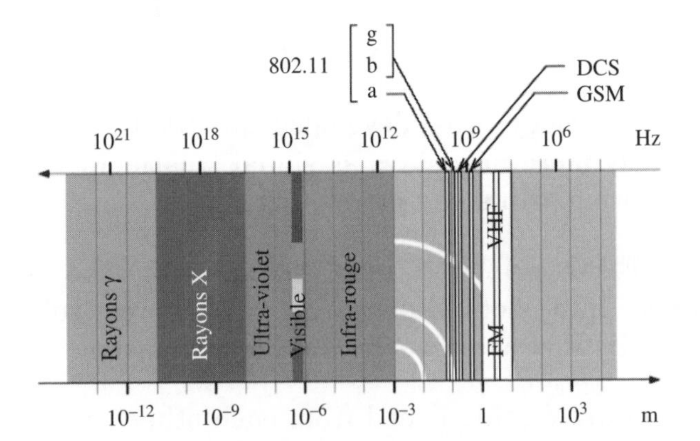

Figure 8.6. *The ISM band*

8.2.2. *The data link layer*

The data link layer of the 802.11 norm is made up of two sublayers:

– the LLC sublayer and

– the MAC sublayer.

The 802.11 standard uses LLC 802.2 and 48 bit address fields, as in all other local area networks (LAN 802s), simplifying connections between wireless and wired networks.

The MAC sublayer of the 802.11 norm is very close to the 802.3 norm. It was conceived to support multiple users over shared media, based on a procedure of listening before any transmission occurs.

For Ethernet 802.3 LANs, machines use the carrier sense multiple access with collision detection (CSMA/CD) access method, in which each machine is free to communicate at any moment. Each machine sending a message checks that no other message has been sent at the same time by another machine. If a simultaneous attempt to send is detected, both machines wait for a random length of time before recommencing transmission.

The CSMA protocol regulates Ethernet stations' access to the cable. It also detects and manages collisions that occur when two or more peripherals attempt to communicate simultaneously over the LAN.

In a wireless environment, this procedure is not possible because two stations communicating with a receiver may not be able to hear each other due to their limited range. To detect a collision, a station must be capable of transmitting and listening at the same time. In a radio system, however, this cannot be done. A new access method with collision detection, CSMA-CA has been developed to manage access to the radio channel in a wireless network.

8.2.2.1. *CSMA / CA*

The CSMA/CA protocol attempts to avoid collisions (two frames sent almost simultaneously which crash) as far as possible by imposing systematic acknowledgment (ACK) of packets. An ACK packet is sent by the receiving station for each data packet received correctly.

The operation of the CSMA/CA protocol (Figure 8.7) can be described as follows:

– A station wishing to send listens to the network.

– If the network is being used, transmission is delayed. If the network is free for a given time, distributed inter frame space (DIFS), then the station begins sending.

– The station transmits a ready to send (RTS) message containing information on the volume of data which it wishes to send and the transmission speed.

– The receiver (an access point (AP) in infrastructure mode) replies with a clear to send (CTS) and the station begins sending data.

– When all the sent data has been received, the receiver sends an ACK.

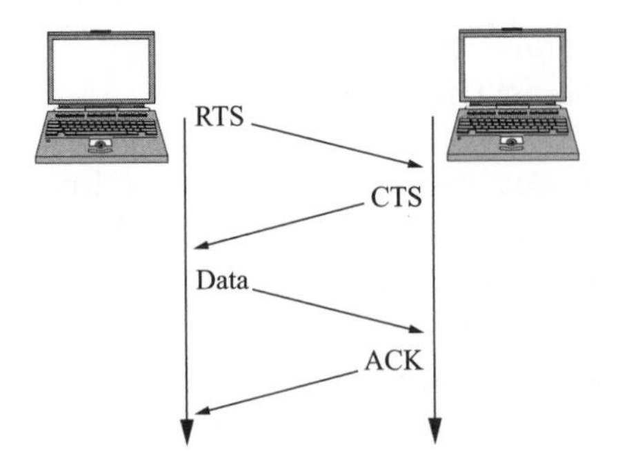

Figure 8.7. *CSMA / CA in a Wi-Fi network*

All neighboring stations wait for the time they consider necessary for the transmission of the given volume of information at the announced speed. If the ACK frame is not detected by the sending station (because either the ACK packet or the original packet has been lost, temporarily or permanently), the nodes assume that a collision has occurred and the data packet is retransmitted after a time delay of random length.

The CSMA/CA protocol allows access to the medium to be shared in a wireless network. The principal anomaly in this mechanism is the addition of extra load to 802.11, unknown in 802.3. We can therefore state that the 802.11 local network will always perform less well than an Ethernet LAN with equivalent theoretical debit.

8.2.2.2. *RTS / CTS*

The use of RTS/CTS frames is intended to combat the problem of "hidden nodes" or "hidden stations," where two stations located on each side of an AP can both hear activity at the AP, but not each other. This problem is usually the result of distance or the presence of an obstacle.

When using the RTS/CTS function, an emitting station sends an RTS message and waits until the AP responds with a CTS message. All stations in the network can hear the AP and the CTS message allows them to manage all planned transmissions.

The station wishing to send can thus send and receive an ACK without risk of collision. The addition of temporary reservation of the support (by the RTS/CTS protocol) to the network load increases the weight of transmissions and can reduce bandwidth.

8.2.2.3 *Modes of access to the wireless channel*

The MAC layer defines two different access methods for the wireless channel: the distributed coordination function (DCF) or contention period (CP), and point coordination function (PCF) or contention free period.

Access to the medium can thus work in two ways depending on the topology and performance of the wireless network.

8.2.2.3.1. DCF mode

The DCF mode of access is similar to the Ethernet, in that it allows the transport of asynchronous data and stations have an equal chance of being able to access the support (Figure 8.8). Ad hoc networks only use this access method, which is based on the CSMA/CA protocol. The spacing values between frames for different wireless norms are given in Table 8.2.

Parameters	802.11a	802.1b(FH)	802.11b(DS)	802.11b(IR)	802.11b (high rate)
Slot time (□s)	9	50	20	8	20
SIFS (μs)	16	28	10	10	10
DIFS(μs)	34	128	50	26	50
EIFS (μs)	92,6	396	364	205 or 193	268 or 364
CWmin (slot time)	15	15	31	63	31
CWmax (slot time)	1023	1,023	1,023	1,023	1,023

Table 8.1. *Frame spacing*

DCF mode with collision avoidance uses RTS/CTS messages following the same approach: packets (of considerable size) cannot be sent until a CTS message has been received (Figure 8.8).

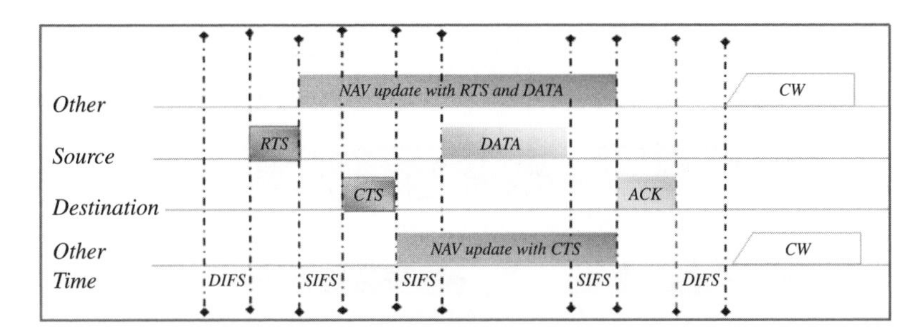

Figure 8.8. *RTS / CTS exchange in DCF mode*

8.2.2.3.2. PCF mode

PCF mode is a function that guarantees transmission at a regular rhythm, allowing synchronization of fluxes (images, sound, or other) or real-time working. It provides an alternative to CSMA/CA using distributed function.

PCF is based on "questioning" stations, or polling, controlled by the AP. A station may not send without authorization from the AP.

This access mode can only be used if the DCF access mode is already in place. The two access modes alternate according to station demand. This method is designed for use with real-time applications (video, voice calls), which necessitate delay management during data transmission.

PCF consists of two main phases:

– polling, where the AP questions each station and constructs a list of stations chosen to transmit; and

– DCF, where the system evolves freely without the AP intervening: a selected and authorized station carries out its transmissions using DCF mode (Figure 8.9).

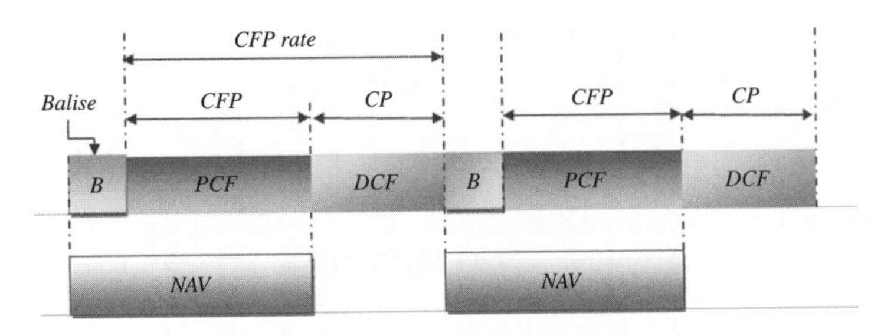

Figure 8.9. *Data exchange in PCF mode*

8.2.2.4. *Additional properties of the MAC and LLC layers*

The MAC layer of 802.11 provides two other characteristics offering robustness: CRC checksums and packet fragmentation. A checksum is calculated for and attached to each packet to ensure that data has not been corrupted during the transfer.

Packet fragmentation allows large packets to be divided up into smaller units, a particularly useful attribute in very congested environments or in case of interference problems, where large packets run a higher risk of corruption. This technique limits the risk of packets needing to be retransmitted and thus improves the global performance of the wireless network.

The MAC layer is responsible for the reconstitution of fragments received; the process is therefore transparent for protocols on the layer above.

The LLC layer of 802.11 also deals with energy management. Two modes of power management are possible:

– continuous aware mode: the unit, always on, consumes energy permanently; and

– power save polling mode: the unit is put onto standby.

The AP then puts data destined for the station on standby into a queue.

8.3. Wi-Fi network architecture

8.3.1. *Ad hoc mode*

Ad hoc mode is an operating mode that allows computers equipped with a Wi-Fi network card to connect directly to each other without using outside material such as an AP. To establish a network of this kind, machines must be configured in an ad hoc mode, a channel (frequency) must be selected and a service set identifier (SSID) network name, shared by all the machines in the network, must be chosen.

The advantage of this mode is that there is no need for costly additional material and that a network of this kind is easy to set up. Using a simple routing program, as described in the previous chapters, it is possible to create autonomous grid networks in which range is not limited to the immediate neighbors of a node (as all participants act as routers).

Ad hoc mode is characterized by its flexibility. An ad hoc network does not require any preexisting infrastructure to enable communication between its members (Figure 8.10). Mobile users communicate with each other directly. Each machine plays a double role of client and gateway to maintain connectivity for itself and other machines.

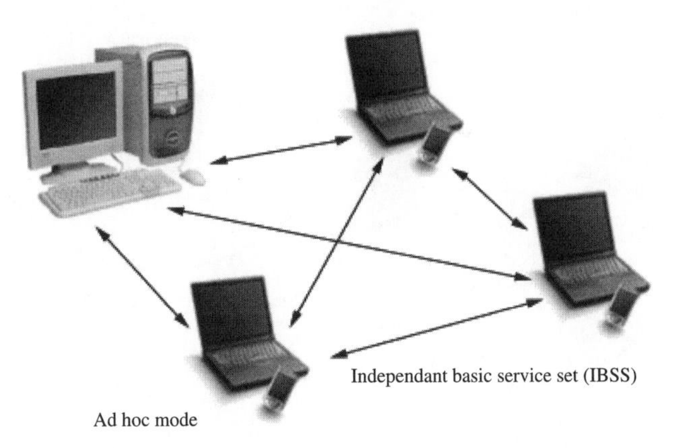

Figure 8.10. *Architecture of an ad hoc network*

The group formed by the different stations is known as an independent basic service set (IBSS). An IBSS is a wireless network constituted of a minimum of two stations capable of communicating between each other without the presence of an AP.

8.3.2. *Infrastructure mode*

Infrastructure mode is an operating mode that allows computers equipped with a Wi-Fi network card to communicate with each other via one or more APs which act as hubs (e.g. hub/switch in a wired network). This mode is mostly used in businesses.

To set up a network of this kind, APs must be placed at regular intervals in the zone to be covered by the network (Figure 8.11). The APs, and the machines involved, must be configured with the same SSID to communicate. The advantage of this mode is that communications must pass through the AP; it is thus possible to control network access.

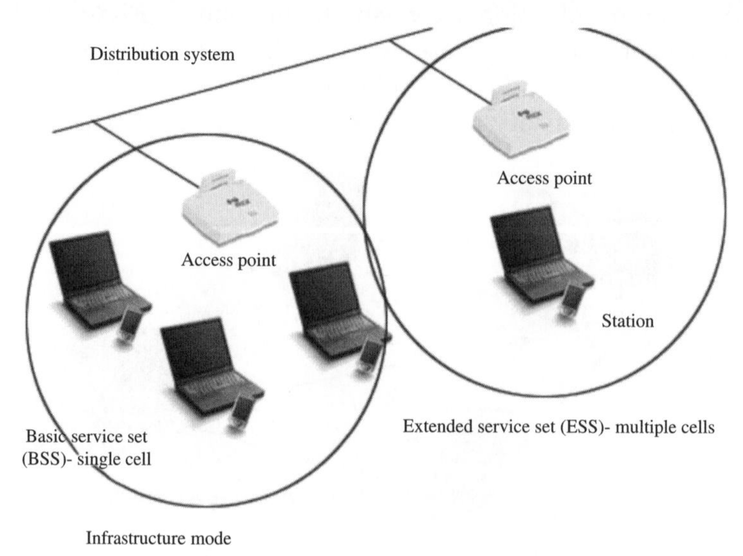

Figure 8.11. *Infrastructure mode*

However, the network cannot be made bigger without installing new hubs. Users, whether fixed or mobile, send all communications to the AP that maintains connectivity with stations in its zone of coverage.

The group formed by the AP and the stations within its zone of coverage is known as a basic service set (BSS) and constitutes a cell.

Each BSS is identified by a basic service set identifier (BSSID) of six bytes (48 bits). In infrastructure mode, the BSSID corresponds to the MAC address of the AP.

8.3.2.1. *Communication with the AP*

When a station enters a cell, it diffuses a probe request message across all channels containing the SSID for which it is configured and the debits supported by its wireless adapter. If no SSID is configured, the station listens to the network and attempts to find an SSID.

Each AP regularly (approx. every 0.1 s) diffuses a beacon message with information on its BSSID, its characteristics and, usually, its SSID. The SSID is sent by default but it is possible to deactivate this option.

For each probe request received, the AP checks the SSID and the debit present in the beacon. If the SSID corresponds to that of the AP, the AP sends a reply containing information on load and synchronization data. The station receiving the reply can then work out the quality of the signal from the AP, allowing it to calculate its distance. Generally, the closer the AP, the better the debit. A station within the range of more than one AP (all with the same SSID) is free to choose the AP with the best compromise between debit and charge (Figure 8.12).

To associate itself with an AP, the client sends an initial frame to declare its presence. The hub sends, in its response, the functionalities it supports. In the case of an installation with multiple APs, the station will choose the best AP for its needs by analyzing the responses received.

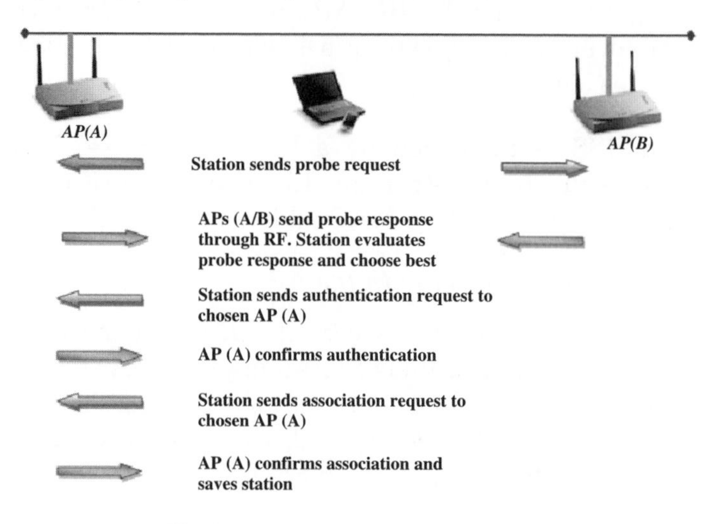

Figure 8.12. *Association process*

The 802.11 authentication requires four exchanges, but uses null authentication. Following the authentication process, the client becomes associated with the AP.

8.3.2.2. *Roaming*

Roaming is the process of movement from one cell towards another without breaking an open connection. This function is similar to handover in cellular systems, with two major differences:

– In a LAN, the transition from one cell to another must be made between two frame transmissions. This is not the case in mobile telephony, where the transition can occur during a conversation.

– In the case of voice transmissions, a temporary disconnection may not affect the conversation, whereas in frame transmission, performance suffers noticeably when lost frames have to be retransmitted.

– Communication between APs involved in roaming is carried out via the inter AP protocol.

– APs must update their MAC address tables to avoid data losses.

Roaming in the 802.11 is a two-step process:

– Listening, actively or passively, to the channel in search of other APs with better signal power

– The reassociation process, where the station is associated with another AP (Figure 8.13).

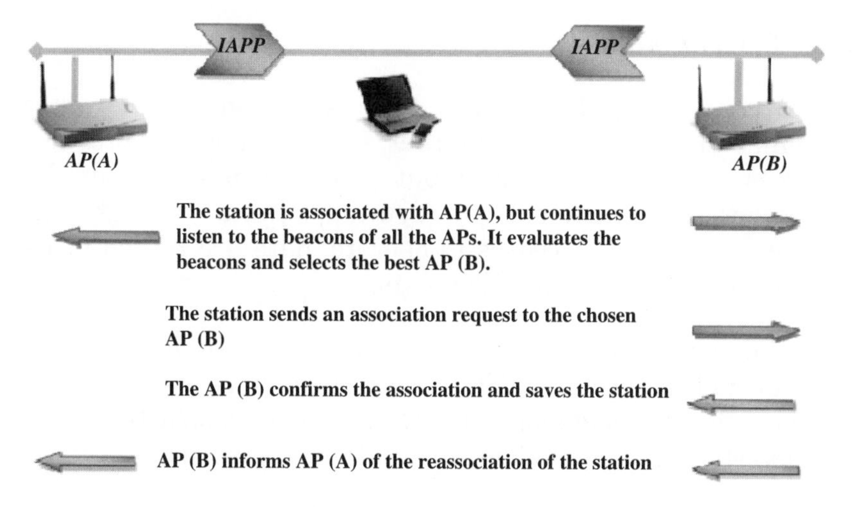

Figure 8.13. *Roaming*

A reassociation of this kind usually occurs when a station moves away from the original AP, making the signal received from the AP weaker. It can also happen as a result of changes in the radio characteristics bit error rate or block error rate, or because of an increase in network traffic at the original AP (overloading). In the latter case, roaming can be used to rebalance load, as it ensures the charge of the WLAN is distributed as efficiently as possibly over the available wireless infrastructure.

This dynamic association/reassociation process at the APs makes it possible to cover a wide geographic area by superposing cells across a block of flats or a campus. If two APs have partially overlapping coverage zones and use overlapping channels, interference may occur which reduces the available bandwidth. DSSS mode only has 14 radio channels of which three are totally isolated, often 1, 6, and 11. These three channels are the best adapted to large-scale multicellular coverage.

8.3.3. *Grid mode*

Multipoint grid networks have a routed grid topology similar to that for wired Internet. Each AP not only provides access for associated used but also becomes part of the network infrastructure by routing traffic through the network. If a point ceases to operate, data are routed to its destination by other (relay) points.

This simply means that the elements of the network are connected to each other and, in practice, more specifically to their closest neighbors. In terms of infrastructure, the network type is equivalent to peer-to-peer networks.

Grid technology enables wireless devices to connect to each other on a peer-to-peer basis, dynamically or statically, instantly, without a central hierarchy. This forms a net-shaped structure, hence the name "grid" (Figure 8.14).

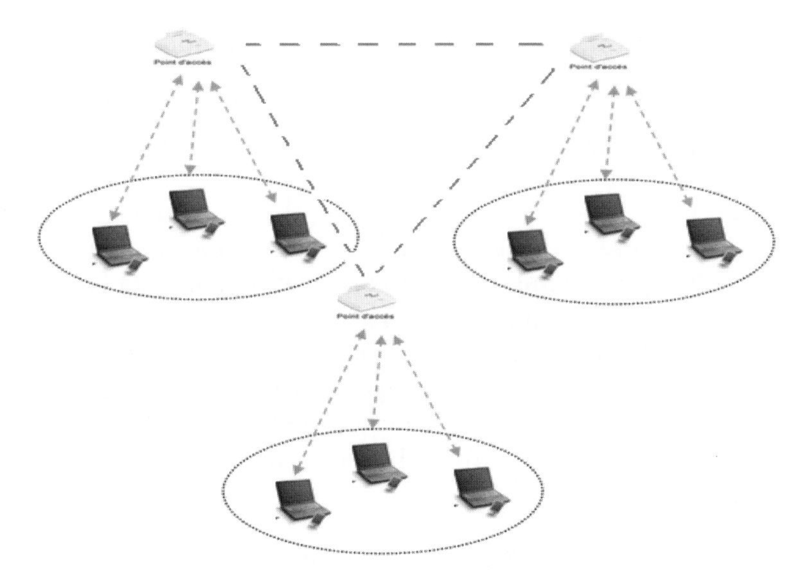

Figure 8.14. *Grid network*

Grid technology also allows new relay stations to connect and disconnect without needing to configure the network manually.

8.4. Wi-Fi norms

The material demands of wireless networks operating at different frequencies and different debits are very important. To respond to these demands, the 802.11 group has published a series of wireless norms.

8.4.1. *The 802.11a norm*

The IEEE 802.11a norm is a variant of Wi-Fi technology which uses a frequency band known as unlicensed national information infrastructure (UNII), subdivided into three sections:

– UNII-1 (from 5.150 to 5.250 GHz), used indoors;

– UNII-2 (from 5.250 to 5.350 GHz), indoor/outdoor; and

– UNII-3 (from 5.725 to 5.825 GHz), used outdoors.

One advantage of this norm is that it solves certain problems present in the 802.11b (in terms of debit and vulnerability) by using different types of modulation (OFDM) and a frequency band, ISM, which is relatively uncluttered. Note that the frequency bands at 2 and 5 GHz are free, i.e. they may be used without license in Europe. Moreover, the theoretical velocity of 54 Mbps is considerably better adapted to the exchange of large files than that of the 802.11b, which offers 11 Mbps.

The 802.11a also has disadvantages, including reduced range and incompatibility with the 802.11b (changing to this norm thus requires entirely new material).

Table 8.3 shows the central frequencies of channels used for the 802.11a norm.

Frequency	Channel number	Transmit frequency	Maximum transmit power (mW)
U-NII lower band	40	5.200	40 mW
	36	5.180	
	44	5.220	
	48	5.240	
U-NII middle band	52	5.260	200 mW
	56	5.280	
	60	5.300	
	64	5.320	
U-NII upper band	149	5.745	800 mW
	153	5.765	
	157	5.785	
	161	5.805	

Table 8.2. *Frequencies and channels used by the 802.11a norm*

8.4.2. *The 802.11b norm*

The term Wi-Fi refers to this norm, the first WLAN norm to pass into wide-scale usage. At the time of writing, the 802.11b has been largely replaced by the 802.11g, which is faster.

The 802.11b norm allows interoperability with existing material. It offers debit of 11 Mbps, with a range of 300 m in an unobstructed environment.

The 802.11b norm is defined by a DSSS modulation, access using CSMA/CA and carrier detection. It uses the free frequency band of 2.4 GHz, subdivided into 13 subchannels of 22 MHz in Europe (11 in the USA and 14 in Japan) with partial overlap (Table 8.4).

Channel	1	2	3	4	5	6	7	8	9	10	11	12	13
Frequency	2.412	2.417	2.422	2.427	2.432	2.437	2.442	2.447	2.452	2.457	2.462	2.467	2.472

Table 8.3. *802.11b frequencies and channels*

In practice, there are only three separate radio channels which do not interfere with each other, namely, channels 1, 6, and 11.

The main disadvantage of the 802.11b is the level of possible interference with other devices functioning over the same frequency, including microwave ovens and wireless analog film cameras. The same frequency is also used by all forms of surveillance, professional or domestic observation, home transmitters, telemetry, telemedicine, amateur radio, wireless keyboards and mice.

8.4.3. *The 802.11e norm*

The 208.11e norm contains modifications to the MAC layer to improve the quality of service (QoS). Communications are planned to occur at moments when no traffic is being transmitted. These optimizations aim to enable Internet protocol telephony and continuous video diffusion.

The norm offers QoS possibilities in the data link layer of the 802.11. It defines the needs of different packets in terms of bandwidth and transmission delay to prioritize those fluxes which require it. Applications of this kind make up a growing market, for example Wi-Fi telephones and Wi-Fi television.

8.4.4. *The 802.11f norm*

The 802.11f norm manages interoperability between APs from different manufacturers. This norm facilitates roaming. The norm is a recommendation to vendors of the 802.11 equipment/tool and aims to improve interoperability between products.

The 802.11f norm allows a mobile user to change APs transparently while moving, independently of the brand or maker of the APs. In reality, makers of the 802.11 equipment/tool sometimes use incompatible proprietary norms that cause problems with roaming.

8.4.5. *The 802.11g norm*

The 802.11g norm was the most widespread Wi-Fi norm in January 2005. An extension of the 802.11b norm increases maximum debit to 54 Mbps using the OFDM modulation technique.

The 802.11g norm also functions at 2.4 GHz, making the two norms perfectly compatible. The 802.11b equipment/tool can therefore be used with the 802.11g APs and vice versa.

Unfortunately, this standard is just as sensitive to interference with other equipment/tool using the same band of frequencies. At the time this standard appeared on the market, users had begun to require increased assurances concerning data security. The 802.11b does not always guarantee security and the proposed encryption method, wired equivalent privacy (WEP), is fairly weak in reality. This problem was dealt with by using Wi-Fi protected access (WPA) instead of WEP.

8.4.6. *The 802.11i norm*

This norm manages authentication and security mechanisms at the MAC layer level. It solves the problems of weakness in the WEP protocol designed for the 802.11 MAC layer. The aim of the 802.11i norm is to improve transmission security (dynamic management and distribution of keys, encryption of information, and user authentication).

The 802.11b and 802.11g norms use WEP to make transmissions secure using encryption keys. The mode of encryption used is Rivest Cypher 4, which has been shown to be weak.

The 802.11i norm offers two complete solutions, including use of the WPA2 algorithm, a much-improved version of WPA. It uses extensible authentication protocol (EAP) authentication, defined in 802.1x, and is based on advanced encryption standard.

Moreover, the norm ensures confidentiality using temporary key integrity protocol encryption, which performs better than the algorithm used with the 802.11g and 802.11b norms. This norm was standardized in June 2004.

8.4.7. *The 802.1x norm*

The 802.1x is a subsection of the 802.11i work group that aims to integrate the EAP protocol. The 802.1x norm deals with secure transmission of information in wired and wireless networks using authentication. The norm supports various methods of authentication, including token cards, single-use passwords, certificates, and public keys.

One application is the use of a remote authentication dial in user service authentication server combined with dynamic distribution of keys, which guarantees a high level of security.

8.4.8. *The 802.11n specification*

This Wi-Fi norm should overcome various problems present in the existing norms, in terms of debit, range, etc. The 802.11n norm was expected to be published in April 2007 but has yet to appear.

This latest 802.11 norm optimizes the debit of the standard, mainly through the use of multiple input multiple output (MIMO) systems that use several frequency bands simultaneously OFDM. This will allow theoretical debits of up to 540 Mbps (real debit around 100 Mbps) and range of around 150 m.

MIMO is a technology that uses several transmission and reception channels simultaneously to transfer spatially coded information through several virtual channels.

The main difference between MIMO and traditional wireless systems resides in the use of the physical phenomenon that is multipath routes, to the benefit of the radio transmission.

Unlike traditional systems, where multipath creates problems, MIMO takes advantage of the phenomenon by generating almost independent signals in the available space, greatly optimizing the transmission system.

8.5. 802.11n migration

The 802.11n variant is an amendment to the existing 802.11 norm, which currently incorporates 80211a/b and g. The new norm requires a new physical layer, with changes to the lower half of the data link layer, particularly the MAC sublayer. The new norm adds more than 500 pages of changes and additions to the current norm, 802.11g (Figure 8.15).

In 802.11n, other aspects of the Wi-Fi network, from the upper half of the data link layer to the application layer, remain the same; in other words, switches, protocols, etc. remain the same.

The main advantage of the new 802.11n norm is in the higher outgoing debit offered to users of MIMO technology, which, in practice, can reach well over 100 Mbps.

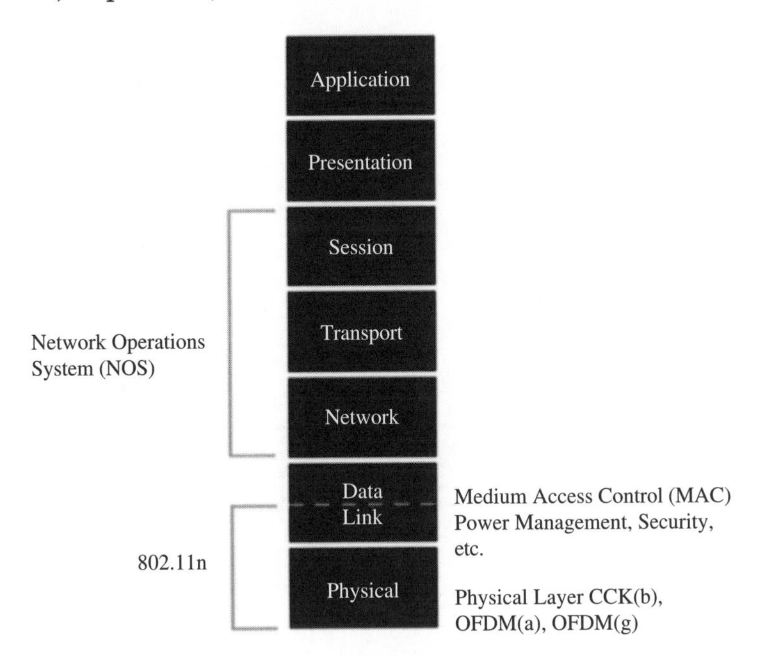

Figure 8.15. *Structure of 802.11n*

The 802.11n will also be retrocompatible, so that the 802.11a/b/g clients will still be able to use their own WLANs without affecting their operation.

Optimal performance, however, can only be produced by an "entirely n" network.

Thus, an update in terms of wireless equipment/tool would be necessary to enable an 802.11n network to demonstrate its full capacities.

As Figure 8.16 shows, debit varies from 11 Mbps with 802.11b to 54 Mbps avec 802.11a and g and can reach hundreds of Mbps using the 802.11n.

In theory, the gross outgoing debit can reach 65 Mbps with a 1×1 configuration (one receiving antenna and one sending antenna) and up to 600 Mbps with a 4×4 configuration.

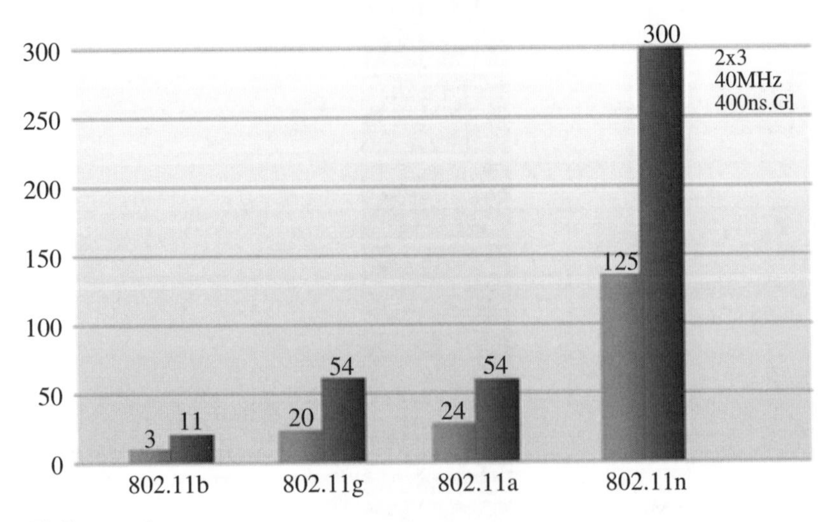

Figure 8.16. *Debit comparison: 802.11a / b / g and n*

8.6. Conclusion

Since the creation of the Wi-Fi norm, permitting interoperability between different existing materials, the installation of this kind of wireless network is becoming more and more necessary. This evolution is the result of a growing demand for mobility from users and the relatively low cost of equipment/tool. Wi-Fi norms have effectively created a revolution in wireless networks. They continue to evolve, and the service currently offered by 802.11g is but a taster of the possibilities of 802.11.

The principal advantage of Wi-Fi rests in the degree to which it has penetrated current markets; a considerable number of users have invested in this technology. Competing technologies, even if they perform better, are faced with a double obstacle: not only the continual improvement of Wi-Fi performance but also its popularity. The main defects of older versions of 802.11 will disappear as the technology progresses. Thus, security, debit, and QoS will be major assets in the next stages of evolution of the 802.11.

The services offered to users will not be limited to a simple connection to the wireless network but also a technically advanced means of communication. The continual progression of communication technology and the "greed" of wireless networks in terms of debit and coverage has led to the appearance of yet another Wi-Fi norm, 802.11n, originally due for publication in 2009.

This latest project aims to extend network coverage even further and increases debit considerably. The 802.11n (still in draft form at the time of writing) allows debit of over 100 Mbps and coverage of a larger zone than its predecessors, using new signaling techniques to achieve these goals.

Bibliography

[AHN 02] AHN G.S., CAMPBELL A.T., VERES A., SUN L.H., "SWAN: service differentiation in stateless wireless ad-hoc networks", *INFOCOM'2002*, New York, June 2002.

[BAD 00] BADIS H., AGHA K.A., "Quality of service for ad hoc optimized link state protocol (QOLSR)", *IETF MANET Draft*, http://tools.ietf.org/wg/manet/draft-badis-manet-qolsr-04.txt, 2000.

[BAS 98] BASAGNI S., CHLAMTAC I., SYROTIUK V.R., WOODWARD B.A., "A distance routing effect algorithm for mobility (DREAM)", *Proc. 4th Annual ACM/IEEE Int. Conf. Mobile Computing and Networking*, ACM Press, p. 76–84, 1998.

[BAS 04] BASAGNI S., CONTI M., GIORDANO S., STOJMENOVI I., *Mobile Ad hoc Networking*, IEEE Press and John Wiley & Sons, New Jersey, New York, 2004.

[BOL 04] BOLENG J., CAMP T., "Adaptive location aided mobile ad hoc network routing", *Proc. 23rd IEEE Int. Performance, Computing, and Communications Conf. (IPCCC'04)*, Phoenix, USA, p. 423–432, 2004.

[BRA 97] BRADEN R., ZHANG L., BERSON S., HERZOG S., JAMIN S., "Resource reservation protocol (RSVP)", *IETF RFC 2205*, September 1997.

[BRO 03] BROWN T.X., DOSHI S., BHANDARE S., "Energy-aware dynamic source routing", IETF Internet Draft, work in progress, http://www.ietf.org/internet-drafts/draft-brown-eadsr-00.txt, June 2003.

[BUR 05] BÜR K., ERSOY C., "Ad hoc quality of service multicast routing", *Elsevier Science Computer Communications*, vol. 29, no. 1, p. 136–148, December 2005.

[CAP 05] CAPONE A., FILIPPINI I., GARCIA DE LA FUENTE M.A., PIZZINIACO L., "SiFT: an efficient method for trajectory based forwarding", *Int. Symposium on Wireless Communication Systems 2005 (ISWCS2005)*, Siena, Italy, September 2005.

[CAR 03] Carpenter B., NICHOLS K., Differentiated Services (DiffServ) Working Group, IETF, http://www.ietf.org/html.charters/OLD/diffserv-charter.html, 2003.

[CIS 07] CISCO, Cisco, Interior Gateway Routing Protocol Documentation, http://www.cisco.com/univercd/cc/td/doc/cisintwk/ito_doc/ igrp.htm, last accessed 02/09/2007.

[CLA 03] CLAUSEN T., JAQUET P., Optimized link state routing protocol (OLSR), Experimental RFC 3626, IETF, http://www.ietf.org/rfc/ rfc3626.txt, October 2003.

[COL 06] COLTUN R., BAKER F., OSPF Version 2 management information base, RFC 4750, IETF, http://www.ietf.org/rfc/rfc4750.txt, December 2006.

[EME 07] EMETHNI M., *Cours* de *théorie des graphes*, Pierre Mendès France University, Grenoble II, http://brassens.upmf-grenoble.fr/IMSS/MathSHS/MASS3/COURS/Chap3.htm, 2007.

[ENC 07a] WEBOPEDIA VIRTUAL ENCYCLOPEDIA, definition of "Load-balancing", http://www.webopedia.com/TERM/l/load_balancing.html, last accessed 02/09/2007.

[ENC 07b] WIKIPEDIA VIRTUAL ENCYCLOPEDIA, Algorithme de Ford-Bellman, http://fr.wikipedia.org/wiki/Algorithme_de_Ford-Bellman, last accessed 02/09/2007.

[GAN 04] GANJALI Y., KESHAVARZIAN A., "Load balancing in ad hoc networks: single-path routing vs. multi-path routing", *Twenty-third Annual Joint Conf. IEEE Computer and Communications Societies (INFOCOM 2004)*, Hong Kong, March 2004.

[GOM 01] GOMEZ J., CAMPBELL A.T., NAGSHINEH M., BISDIKIAN C., WATSON T.J., Power-aware routing optimization protocol (PARO), Internet Draft, work in progress, http//comet.ctr.columbia.edu/~javierg/paro/draft-gomez-paro-manet-00.txt, June 2001.

[HAA 02] Haas Z.J., PEARLMAN M.R., SAMAR P., The zone routing protocol (ZRP) for ad hoc networks, Internet Draft (expired), http://tools.ietf.org/html/draft-ietf-manet-zone-zrp-04, July 2002.

[HED 88] HEDRICK C., Routing information protocol, RFC 1058, IETF, http://www.ietf.org/rfc/rfc1058.txt, June 1988.

[ISL 07] WIKIPEDIA VIRTUAL ENCYCLOPEDIA, Inverse-Square Law, http://en.wikipedia.org/wiki/Inverse_square, last accessed 02/09/2007.

[JIA 99] JIANG M., LI J., TAY Y.C., Cluster based routing protocol (CBRP), IETF Internet Draft, http://tools.ietf.org/html/draft-ietf-manet-cbrp-spec-01, August 1999.

[JOH 03] JOHNSON D.B., MALTZ D.A., HU Y.C., The dynamic source routing protocol for mobile ad hoc networks (DSR), IETF Internet Draft (expired), http://www3.ietf.org/proceedings/04mar/I-D/draft-ietf-manet-dsr-09.txt, April 2003.

[KO 98] KO Y.B., VAIDYA N.H., "Location-aided routing in mobile ad hoc networks", *Proc. ACM/IEEE Mobicom Conf.*, p. 66–75, October 1998.

[LEE 00] LEE S.B., Ahn G.S., ZHANG X., CAMPBELL A.T., "INSIGNIA: An IP-based quality of service framework for mobile ad hoc networks", *Journal of Parallel and Distributed Computing*, vol. 60, no. 4, p. 374–406, 2000.

[LEE 05] LEE Y.J., RILEY G.F., "A workload-based adaptive load-balancing technique for mobile ad hoc networks", *IEEE Wireless Communications and Networking Conference (WCNC' 2005)*, vol. 1, p. 2002–2007, 2005.

[LIN 02] Lindgren A., SCHELEN O., "Infrastructured ad hoc networks", p. 64–70, August 2002.

[LU 03] LU H.L., FAYNBERG I., "An architectural framework for support of quality of service in packet networks", *IEEE Communications Magazine*, vol. 41, no. 6, p. 98, June 2003.

[NGU 06] NGUYEN D.Q., MINET P., "QoS support and OLSR routing in a mobile ad hoc network", *Proc. Int. Conf. Networking, Int. Conf. Systems and Int. Conf. Mobile Communications and Learning Technologies (ICNICONSMCL'06)*, p. 74, April 2006.

[NS 07] NS-2 PROJECT, http://www.isi.edu/nsnam/ns/, last accessed 02/09/2007.

[PAR 06] PARKER J., Management information base for intermediate system to intermediate system (IS-IS), RFC 4444, IETF, http://www.ietf.org/rfc/rfc4444.txt, April 2006.

[PER 94] PERKINS C., "Highly dynamic destination-sequenced distance-vector routing (DSDV) for mobile computers", *ACM SIGCOMM'94 Conf. Communications Architectures, Protocols and Applications*, p. 234–244, 1994.

[PER 01] PERKINS C.E., Royer E.M., Quality of service for ad hoc on-demand distance vector routing, IETF draft, http://tools.ietf.org/html/draft-perkins-manet-aodvqos-00, November 2001.

[PER 03] PERKINS C., ROYER E.M., DAS S., Ad hoc on-demand distance vector (AODV) routing, Experimental RFC 3561, IETF, www.ietf.org/rfc/rfc3561.txt, July 2003.

[PER 00] PERLMAN M., HAAS Z., SCHOLANDER P., TABRIZI S., "Alternate path routing for load balancing in mobile ad hoc networks", *IEEE Military Communications Conf. (MILCOM 2000)*, October 2000.

[PHA 02] PHAM P., PERREAU S., "Multi-path routing protocol with load balancing policy in mobile ad hoc networks", *IFIP Int'l Conf. Mobile and Wireless Communications Networks (MWCN 2002)*, September 2002.

[RIC 89] RICINIELLO F., "The CCITT and the quality of service", *IEEE Int. Conf. Communications*, vol. 1, p. 389–393, 11–14 June 1989.

[ROY 02] ROY S., BANDYOPADHYAY S., UEDA T., HASUIKE K., "Multipath routing in ad hoc wireless networks with omni directional and directional antenna: a comparative study", *Proc. 4th Int. Workshop on Distributed Computing, Mobile and Wireless Computing (IWDC)*, p. 184–191, 2002.

[ROY 99] ROYER E.M., PERKINS C.E., "Multicast operation of the ad-hoc on-demand distance vector routing protocol", *Proc. 5th Annual ACM/IEEE Int. Conf. Mobile Computing and Networking* (*MobiCom*), p. 207–218, August 1999.

[SCH 02] SCHMID A., STEIGNER C., "Avoiding counting to infinity in distance vector routing", *Telecommunication Systems*, vol. 19, no. 3–4, March 2002.

[TAB 97] TABBANE S., *Réseaux mobiles*, Hermes, Paris, p. 290, 1997.

[VAL 03] VALERA A., SEAH W., RAO S.V., "Cooperative packet caching and shortest multipath routing in mobile ad hoc networks", *Proc. IEEE INFOCOM*, 2003.

[WAL 91] WALDBUSSER S., Appletalk management information base, RFC 1243, IETF, http://tools.ietf.org/html/rfc1243, July 1991.

[WAN 00] WANG L., ZHANG L.F., SHU Y.T., DONG M., YANG O.W.W., "Multipath source routing in wireless ad hoc networks", *Proc. IEEE CCECE*, p. 479, 2000.

[WAN 07] WANG X., Implémentation de l'algorithme de Dijkstra sous Matlab, http://www.mathworks.com/matlabcentral/files/5550/dijkstra.m, 2007.

[WEI 07] WEISSTEIN E.W., Graph center of mathworld website, Wolfram Web Resource, http://mathworld.wolfram.com/GraphCenter.html, last accessed 02/09/2007.

[WRO 00] WROCLAWSKI J., Integrated services (charter) charter, IETF, http://www.ietf.org/html.charters/OLD/intserv-charter.html, 2000.

[YIN 04] YIN S., LIN X., "MALB: MANET adaptive load balancing", *IEEE Vehicular Technology Conference (VTC2004-Fall)*, vol. 4, p. 2843–2847, September 2004.

[ZHA 05] ZHANG Z., MA M., YANG Y., "Energy efficient multi-hop polling in clusters of two-layered heterogeneous sensor networks", *Proc. 19th IEEE Int. Parallel and Distributed Processing Symposium*, Denver, Colorado, p. 81b, April 2005.

[ZHO 01] ZHOU A., HASSANEIN H., "Load-balanced wireless ad hoc routing", *IEEE Canadian Conference on Electrical and Computer Engineering*, vol. 2, p.1157–1161, 2001.

APPENDICES

Appendix 1

The Ad Hoc Networks Simulator (ANS)

A1.1. The ANS Application

To carry out our simulations, we contributed to the modification and improvement of an application by adding the following models:

– TCL script generation for the OLSR protocol and simulation of various possible scenarios.

– Support of new the OLSRQSUP protocol integrated at ns level, with script generation and execution of simulations.

– Integration of nam file generation procedure (the existing application did not allow this).

– The ANS application is compatible with version ns-2.27 and version 2.28 (Figures A1.1–A1.3).

However, for the application to be supported and useable in version 2.28, some modifications to the file /mac/channel.cc are needed.

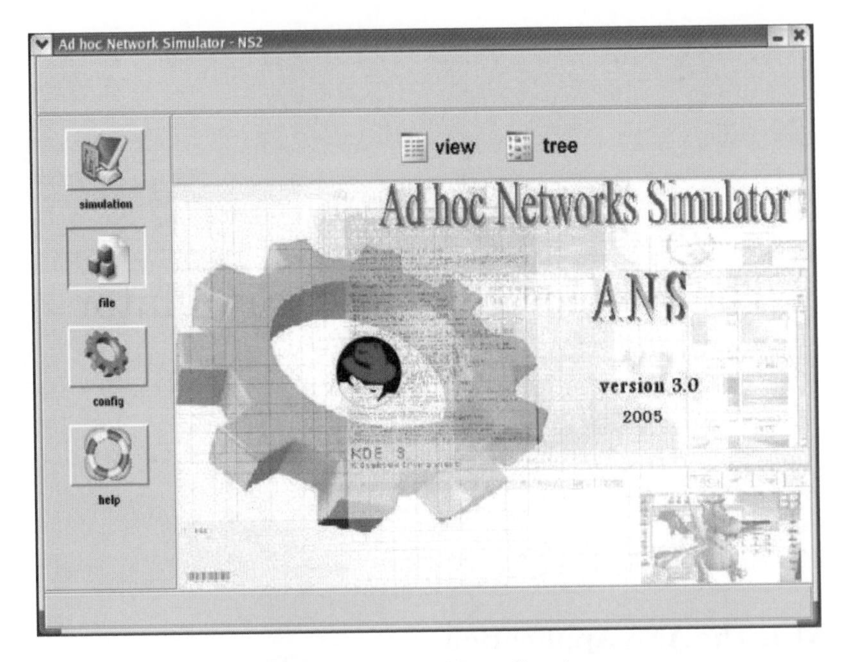

Figure A1.1. *ANS application*

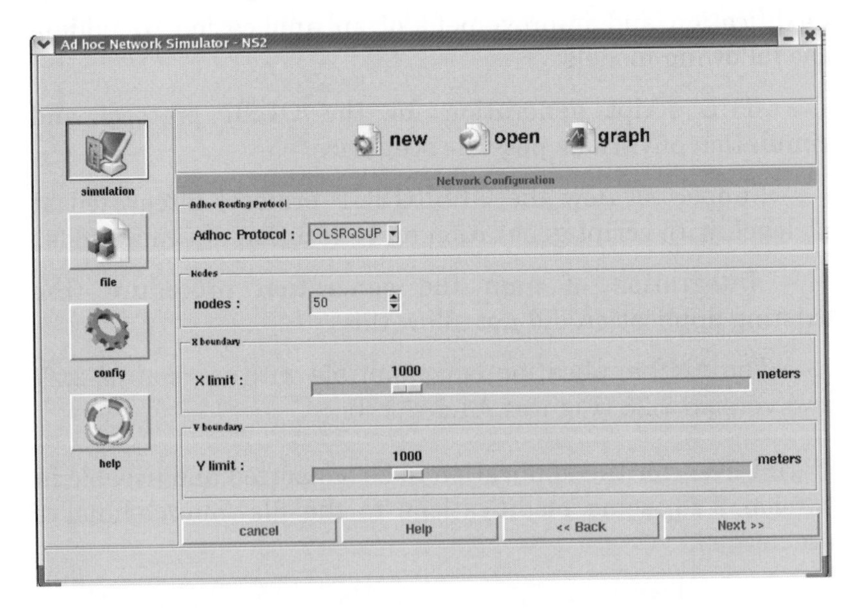

Figure A1.2. *ANS support for OLSRQSUP protocol*

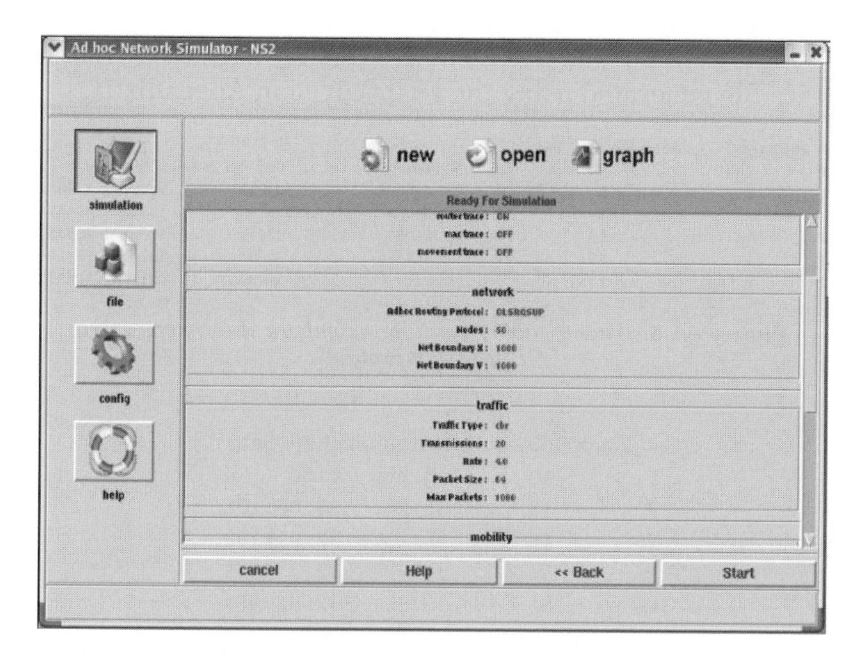

Figure A1.3. *Simulation of an OLSRQSUP protocol scenario using ANS*

A1.2. Traces

We added functions that allow tables for the (OLSR/OLSRQSUP) protocol to be established at trace-file level by referring to the script level.

Figure A1.4 shows an extract from a trace file using the OLSRQSUP protocol (Figure A1.5).

```
r -t 9.295831216 -Hs 14 -Hd -1 -Ni 14 -Nx 441.03 -Ny 1.63 -Nz 0.00 -Ne -1.000000 -Nl RTR -Nw --- -Ma 0 -Md ffffffff -Ms 0 -Mt 800 -Is
0.255 -Id -1.255 -It OLSRQSUP -Il 136 -If 0 -Ii 134 -Iv 32 -P olsrqsup -Pn 1 -Ps 5 [-Pt HELLO -Po 0 -Ph 0 -Pms 5]
s -t 9.488238850 -Hs 6 -Hd -1 -Ni 6 -Nx 241.27 -Ny 31.94 -Nz 0.00 -Ne -1.000000 -Nl RTR -Nw --- -Ma 0 -Md 0 -Ms 0 -Mt 0 -Is 6.255 -
Id -1.255 -It OLSRQSUP -Il 68 -If 0 -Ii 135 -Iv 32 -P olsrqsup -Pn 1 -Ps 6 [-Pt TC -Po 6 -Ph 0 -Pms 6]
```

Figure A1.4. *Extract from trace file using OLSRQSUP*

```
r -t 9.295767216 -Hs 14 -Hd -1 -Ni 14 -Nx 441.03 -Ny 1.63 -Nz 0.00 -Ne -1.000000 -Nl RTR -Nw --- -Ma 0 -Md ffffffff -Ms 0 -Mt 800 -Is
0.255 -Id -1.255 -It OLSR -Il 128 -If 0 -Ii 134 -Iv 32 -P olsr -Pn 1 -Ps 5 [-Pt HELLO -Po 0 -Ph 0 -Pms 5]
s -t 9.488238850 -Hs 6 -Hd -1 -Ni 6 -Nx 241.27 -Ny 31.94 -Nz 0.00 -Ne -1.000000 -Nl RTR -Nw --- -Ma 0 -Md 0 -Ms 0 -Mt 0 -Is 6.255 -
Id -1.255 -It OLSR -Il 66 -If 0 -Ii 135 -Iv 32 -P olsr -Pn 1 -Ps 6 [-Pt TC -Po 6 -Ph 0 -Pms 6]
```

Figure A1.5. *Extract from trace file using OLSR*

Figures A1.6 and A1.7 show a listing extracted from a trace file generated using the script given in Appendix II.

```
P 28.000000 _3_ Neighbor2hop Set
P    Nb    Nb_2hop    N_time       N_2hop_delay    N_2hop_Bw
P    0     9          33.547065    5.803332        1982142.112088
P    18    5          32.590892    3.219186        1960014.806354
P    18    13         32.590892    5.742013        1960191.806354
P    5     18         33.396370    3.883866        1978686.526196
P    5     13         33.396370    5.551286        2012663.526196
```

Figure A1.6. *Trace of table format for neighbors at two hops using OLSRQSUP protocol*

```
P 28.000000 _3_ Neighbor2hop Set
P    nb    nb2hop    time
P    18    5         32.601101
P    18    13        32.601101
P    5     18        32.326954
P    5     13        32.326954
```

Figure A1.7. *Trace of table format for neighbors at two hops using OLSR protocol*

A1.3. Curve tracing

Curve tracing was one of the most important steps in the development phase of this project, as, without the results of the simulations, no conclusions can be reached on the contributions made or on the engineering work carried out.

The following script was written with the aim of being able to plot several curves on the same graph (Figure A1.8).

```
set terminal png
set output "toh.jpg"
show output
set title "Trafic OverHead"
set xlabel "Speed(m/s)"
set ylabel "TOH(pkts)"
set xrange [0:20]
```

```
set yrange [0:20000]
set pointsize 2
set key left top
set xtics 0,5,20
set ytics 0,5000,20000

plot "olsr10.tr" every 1 using 1:2 title "OLSR-10" with linespoints pt 5,
"olsr30.tr" every 1 using 1:2 title "OLSR-30" with linespoints pt 1,
"aodv10.tr" every 1 using 1:2 title "AODV-10" with linespoints, "aodv30.tr"
every 1 using 1:2 title "AODV-30" with linespoints 3
```

Figure A1.8. *Format of courbe.gp file*

Later, the execution of the script using Gnuplot (or even Xgraph) allows curves to be plotted:

```
#gnuplot courbe.gp
```

However, if we wish to use Xgraph, as available with all versions of the simulator, we execute the following command:

```
#exec xgraph courbe.xg
```

Appendix 2

TCL Script of OLSRQSUP Protocol

```
#--------------------PFE 2005--------------------------------------
#This script was written by MAAMER MANEL
#in purpose of simulating OLSRQSUP on ns-2.27
#------------------------------------------------------------------
#==================================================================
# Define options
#==================================================================
set opt(chan)     Channel/WirelessChannel;       # channel type
set opt(prop)     Propagation/TwoRayGround;      # radio-propagation
                                                   model
set opt(netif)    Phy/WirelessPhy;               # network interface
                                                   type
set opt(mac)      Mac/802_11;                    # MAC type
set opt(ifq)      Queue/DropTail/PriQueue;       # interface queue
                                                   type
set opt(ll)       LL;                            # link layer type
set opt(ant)      Antenna/OmniAntenna;           # antenna model
set opt(ifqlen)   50;                            # max packet in ifq
set opt(nn)       50;                            # number of
                                                   mobilenodes
set opt(adhocRouting)  OLSRQSUP;                 # routing protocol
```

```
set opt(cp)      "/root/scenario/traf_50_15_1000";       # connection
                                                          pattern file
set opt(sc)      "/root/scenario/mvn_50_5";      # node movement file.

set opt(x)       1000;              # x coordinate of topology
set opt(y)       1000;              # y coordinate of topology
set opt(seed)      0.0;             # seed for random number gen.
set opt(stop)     300;              # time to stop simulation

set opt(cbr-start)   30.0;          #time of starting trafic emission
set opt(time)              10
#==========================================================
#
# check for random seed
#
if {$opt(seed) > 0} {
  puts "Seeding Random number generator with $opt(seed)\n"
  ns-random $opt(seed)
}
#
# create simulator instance
#
set ns_ [new Simulator]

#
# control OLSRQSUP behaviour from this script -
# commented lines are not needed because
# those are default values in ns-default.tcl
# which are the same as in RFC 3626
# but if we need to change them
# we have just to act here
#
#Agent/OLSRQSUP set use_mac_  true
#Agent/OLSRQSUP set debug_   false
#Agent/OLSRQSUP set willingness 3
```

```
#Agent/OLSRQSUP set hello_ival_ 2
#Agent/OLSRQSUP set tc_ival_  5

#Agent/OLSRQSUP set bw_ival_  0.5 #we compute avail bw each
0.5s (QoS)

$ns_ use-newtrace
set tracefd [open out_OLSRQSUP.tr w]
set namtrace [open out_OLSRQSUP.nam w]
$ns_ trace-all $tracefd
$ns_ namtrace-all-wireless $namtrace $opt(x) $opt(y)

#
# create topography object
#
set topo [new Topography]

#
# define topology
#
$topo load_flatgrid $opt(x) $opt(y)

#
# create God
#
set god_ [create-god $opt(nn)]

#
# configure mobile nodes
#
$ns_ node-config -adhocRouting $opt(adhocRouting) \
        -llType $opt(ll) \
        -macType $opt(mac) \
        -ifqType $opt(ifq) \
        -ifqLen $opt(ifqlen) \
```

```
        -antType $opt(ant) \
        -propType $opt(prop) \
        -phyType $opt(netif) \
        -channelType $opt(chan) \
        -topoInstance $topo \
        -wiredRouting OFF \
        -agentTrace ON \
        -routerTrace ON \
        -macTrace OFF

for {set i 0} {$i < $opt(nn)} {incr i} {
  set node_($i) [$ns_ node]
}
#
# print (in the trace file) routing table and other
# internal data structures on a per-node basis
#
for {set i 0} {$i < $opt(nn)} {incr i} {

$ns_ at 28 "[$node_($i) agent 255] print_rtable"
$ns_ at 28 "[$node_($i) agent 255] print_linkset"
$ns_ at 28 "[$node_($i) agent 255] print_nbset"
$ns_ at 28 "[$node_($i) agent 255] print_nb2hopset"
$ns_ at 28 "[$node_($i) agent 255] print_mprset"
$ns_ at 28 "[$node_($i) agent 255] print_mprselset"
$ns_ at 28 "[$node_($i) agent 255] print_topologyset"
}

#
# source connection-pattern and node-movement scripts
#
if { $opt(cp) == "" } {
  puts "*** NOTE: no connection pattern specified."
  set opt(cp) "none"
} else {
```

```
 puts "Loading connection pattern..."
 source $opt(cp)
}
if { $opt(sc) == "" } {
 puts "*** NOTE: no scenario file specified."
 set opt(sc) "none"
} else {
 puts "Loading scenario file..."
 source $opt(sc)
 puts "Load complete..."
}

#
# define initial node position in nam
#
for {set i 0} {$i < $opt(nn)} {incr i} {
 $ns_ initial_node_pos $node_($i) 20
}

#
# tell all nodes when the simulation ends
#
for {set i 0} {$i < $opt(nn) } {incr i} {
 $ns_ at $opt(stop).0 "$node_($i) reset";
}

$ns_ at $opt(stop).0002 "puts \"NS EXITING...\" ; $ns_ halt"
$ns_ at $opt(stop).0001 "stop"

proc stop {} {
 global ns_ tracefd namtrace
 $ns_ flush-trace
 close $tracefd
 close $namtrace
}
```

```
#
# begin simulation
#
puts "Starting Simulation..."

$ns_ run
```

Appendix 3

Awk Script

```
#------------------------------------------------------------------------------------------
#Awk file used to filter trace file from olsr simulation Sup'Com
# May 2005,
#maamermanel@yahoo.fr
#We would like to compute the number of MPRs in a network
# using OLSR and OLSRQSUP protocol so we have to know the
# node which sends TC messages since only the MPR brodcast
# such control messages
#MAAMER MANEL
#------------------------------------------------------------------------------------------
BEGIN {
mpr = 0;
i=0;
comp=0;
chrono=0;
k=0;
average=0;
}
($1 == "s" && $19 == "RTR" && $51 == "TC") { mpr++;
        if (i==0) {MPR_id[0]=$9;
                comp=0;
```

```
                k=1;
                i=1;}
        else {
                found = "false";
                for(j=0;j<=comp;j++)
                {
                if (MPR_id[j] == $9) {found="true";}
                }
                if(found != "true"){MPR_id[comp+1]=$9;
                                comp++;}
                if ( ($3 - chrono) >= 5) {
                                total_mpr[k]=comp+1;
                                duree_mpr[k]=5*k;
                                #printf(" %-6d\n",total_mpr[k]);
                                chrono= $3;
                                k++;

                                max_k++;
                                comp=0;
                                }

}
}

END {
#printf("TC messages sent: %-6d\n", mpr);
#printf("The number of MPRs for this network is : %-6d\n", comp+1);
#printf("The MPR nodes for this network are: \n");
#for(j=0;j<=comp;j++)
#{
#printf("Node %-6d \n", MPR_id[j]);
#
#}
for(j=1;j<=max_k;j++)
{
```

```
printf( " %-6d %-6d\n", duree_mpr[j],total_mpr[j]);
average = total_mpr[j]+ average;

}
printf("la moyenne est %-7.6f\n", average/max_k);
printf(" %-6d %\n",max_k);
}
```

Index

L

M

N

P

Q

R

S

W